Study Guide
to
For All Practical Purposes
Seventh Edition

Heidi A. Howard
Florida Community College at Jacksonville

W. H. Freeman and Company
New York

ISBN: 0-7167-6946-8
EAN: 9780716769460

First printing

W. H. Freeman and Company
41 Madison Avenue
New York, NY 10010
Houndmills, Basingstoke
RG21 6XS England

Student Study Guide
Table of Contents

PART I: Management Sciences

Chapter 1   Urban Services
Chapter Objectives.. 1
Guided Reading... 2
Homework Help.. 10
Do You Know the Terms? (flashcards)... 11
Practice Quiz... 15
Word Search.. 17

Chapter 2   Business Efficiency
Chapter Objectives.. 19
Guided Reading... 20
Homework Help.. 28
Do You Know the Terms? (flashcards)... 29
Practice Quiz... 35
Word Search.. 39

Chapter 3   Planning and Scheduling
Chapter Objectives.. 41
Guided Reading... 42
Homework Help.. 51
Do You Know the Terms? (flashcards)... 53
Practice Quiz... 59
Word Search.. 63

Chapter 4   Linear Programming
Chapter Objectives.. 65
Guided Reading... 66
Homework Help.. 78
Do You Know the Terms? (flashcards)... 81
Learning the Calculator... 87
Practice Quiz... 89
Word Search.. 91

PART II: Statistics: The Science of Data

Chapter 5   Exploring Data: Distributions
Chapter Objectives.. 93
Guided Reading... 94
Homework Help.. 105
Do You Know the Terms? (flashcards)... 115
Learning the Calculator... 119
Practice Quiz... 123
Word Search.. 125

Chapter 6   Exploring Data: Relationships
Chapter Objectives.. 127
Guided Reading... 127
Homework Help.. 133
Do You Know the Terms? (flashcards).............................. 139
Learning the Calculator.. 141
Practice Quiz... 145
Word Search.. 147

Chapter 7   Data for Decisions
Chapter Objectives.. 149
Guided Reading... 150
Homework Help.. 160
Do You Know the Terms? (flashcards).............................. 165
Learning the Calculator.. 171
Practice Quiz... 173
Word Search.. 175

Chapter 8   Probability: The Mathematics of Chance
Chapter Objectives.. 177
Guided Reading... 178
Homework Help.. 187
Do You Know the Terms? (flashcards).............................. 195
Practice Quiz... 201
Word Search.. 203

Part III: Voting and Social Choice

Chapter 9   Social Choice: The Impossible Dream
Chapter Objectives.. 205
Guided Reading... 206
Homework Help.. 217
Do You Know the Terms? (flashcards).............................. 223
Practice Quiz... 227
Word Search.. 231

Chapter 10   The Manipulability of Voting Systems
Chapter Objectives.. 233
Guided Reading... 234
Homework Help.. 245
Do You Know the Terms? (flashcards).............................. 255
Practice Quiz... 259
Word Search.. 263

Chapter 11   Weighted Voting Systems
Chapter Objectives.. 265
Guided Reading... 266
Homework Help.. 276
Do You Know the Terms? (flashcards).............................. 283
Practice Quiz... 289
Word Search.. 291

Chapter 12 Electing the President
        Chapter Objectives.. 293
        Guided Reading... 294
        Homework Help... 299
        Do You Know the Terms? (flashcards).. 301
        Practice Quiz... 309
        Word Search.. 311

PART IV: Fairness and Game Theory

Chapter 13 Fair Division
        Chapter Objectives.. 313
        Guided Reading... 314
        Homework Help... 321
        Do You Know the Terms? (flashcards).. 323
        Practice Quiz... 327
        Word Search.. 329

Chapter 14 Apportionment
        Chapter Objectives.. 331
        Guided Reading... 332
        Homework Help... 338
        Do You Know the Terms? (flashcards).. 339
        Practice Quiz... 347
        Word Search.. 349

Chapter 15 Game Theory
        Chapter Objectives.. 351
        Guided Reading... 352
        Homework Help... 360
        Do You Know the Terms? (flashcards).. 361
        Practice Quiz... 371
        Word Search.. 373

Part V: The Digital Revolution

Chapter 16 Identification Numbers
        Chapter Objectives.. 375
        Guided Reading... 376
        Homework Help... 381
        Do You Know the Terms? (flashcards).. 383
        Practice Quiz... 387
        Word Search.. 389

Chapter 17 Information Science
        Chapter Objectives.. 391
        Guided Reading... 392
        Homework Help... 400
        Do You Know the Terms? (flashcards).. 409
        Practice Quiz... 415
        Word Search.. 417

Part VI: On Size and Growth

Chapter 18 Growth and Form
Chapter Objectives……………………………………………………..….   419
Guided Reading………………………………………………………….   420
Homework Help…………………………………………………………   425
Do You Know the Terms? (flashcards)………………………………….   427
Practice Quiz……………………………………………………………   433
Word Search……………………………………………………………..   435

Chapter 19 Symmetry and Patterns
Chapter Objectives…………………………………………………..…...   437
Guided Reading………………………………………………………….   437
Homework Help…………………………………………………….......   444
Do You Know the Terms? (flashcards)………………………………….   445
Practice Quiz……………………………………………………………   451
Word Search……………………………………………………………..   453

Chapter 20 Tilings
Chapter Objectives……………………………………………..………   455
Guided Reading………………………………………………………….   456
Homework Help…………………………………………………….......   462
Do You Know the Terms? (flashcards)………………………………….   463
Practice Quiz……………………………………………………………   469
Word Search……………………………………………………………..   471

Part VII: Your Money and Resources

Chapter 21 Savings Models
Chapter Objectives…………………………………………………..…...   473
Guided Reading………………………………………………………….   474
Homework Help…………………………………………………….......   480
Do You Know the Terms? (flashcards)………………………………….   481
Practice Quiz……………………………………………………………   487
Word Search……………………………………………………………..   489

Chapter 22 Borrowing Models
Chapter Objectives……………………………………………………   491
Guided Reading………………………………………………………….   491
Homework Help……………………………………………...…..........   496
Do You Know the Terms? (flashcards)………………………………….   497
Practice Quiz……………………………………………………………   503
Word Search……………………………………………………………..   505

Chapter 23 The Economics of Resources
Chapter Objectives…………………………………………………..…...   507
Guided Reading………………………………………………………….   507
Homework Help…………………………………………...…..........   512
Do You Know the Terms? (flashcards)………………………………….   513
Practice Quiz……………………………………………………………   521
Word Search……………………………………………………………..   523

Practice Quiz Answers…………………………………………………….   525

Chapter 1
Urban Services

Chapter Objectives

Check off these skills when you feel that you have mastered them.

☐ Determine by observation if a graph is connected.

☐ Identify vertices and edges of a given graph.

☐ Construct the graph of a given street network.

☐ Determine by observation the valence of each vertex of a graph.

☐ Define an Euler circuit.

☐ List the two conditions for the existence of an Euler circuit.

☐ Determine whether a graph contains an Euler circuit.

☐ If a graph contains an Euler circuit, list one such circuit by identifying the order of vertices in the circuit's path.

☐ If a graph does not contain an Euler circuit, add a minimum number of edges to eulerize the graph.

☐ Identify management science problems whose solutions involve Euler circuits.

Guided Reading

Introduction

The management of a large and complex system requires careful planning and problem solving. In this chapter, we focus on an important management issue, one that occurs frequently in a variety of forms: the problem of traversing the network as efficiently and with as little redundancy as possible. Solutions involve mathematical principles as well as practical considerations.

Section 1.1   Euler Circuits

⌥ Key idea

The problem of finding an optimal route for checking parking meters or delivering mail can be modeled abstractly as finding a best path through a **graph** that includes every edge. A graph is a finite set of dots and connecting links. A dot is called a **vertex**, and the link between two vertices is called an **edge**.

⌥ Key idea

The problem of finding an optimal route for checking parking meters or delivering mail can be modeled abstractly as finding a best path through a graph that includes every edge.

ᘓ Example A

Represent the street network of stores to be serviced for delivery as a graph. (The ★ represents a store.)

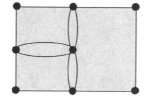

Solution

Start with the basic street network. Without the stores it looks like this:

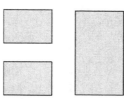

Now replace each intersection or corner with a vertex. Represent these with circles like this:

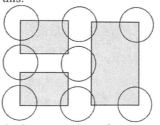

By replacing each row of stores with an edge, the graph in the answer is made.

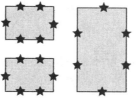

☞ Key idea

A path through a graph is a **circuit** if it starts and ends at the same vertex. A circuit is an **Euler circuit** if it covers each edge exactly once. (Euler is pronounced like "*Oy'lur*")

ᨆ Example B

Draw an Euler circuit of the graph for the store network.

Solution

There are many correct answers.

Starting and ending with the upper left-hand corner, this circuit covers each edge exactly once. Any circuit drawn that meets these conditions (1. starts and ends at the same vertex, and 2. covers each edge exactly once) is an Euler circuit. This is one possible Euler circuit for the graph.

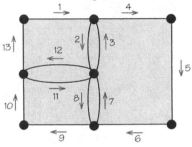

✎ Question 1

Given the following graph, which of the following is true?

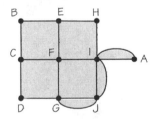

a) An Euler circuit can be found.
b) A circuit can be found if you are allowed to repeat using edges.

Answer

b

Section 1.2   Finding Euler Circuits

☞ Key idea

The **valence** of a vertex is the number of edges that meet at that vertex. This will be either an even or odd positive integer. If the vertex is isolated, then it will have valence zero.

☞ Key idea

A graph is **connected** if for every pair of vertices there is at least one path connecting these two vertices. There are no vertices with valence zero.

⊶ Key idea

According to **Euler's theorem**, a connected graph has an Euler circuit if the valence at each vertex is an even number. If any vertex has an odd valence, there cannot be an Euler circuit.

ᏻ Example C

For each of these graphs, find the valence of each vertex. Which graph has an Euler circuit?

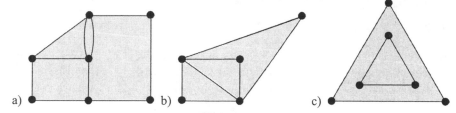

Solution

a)   This graph has two odd vertices, so it cannot have an Euler circuit.

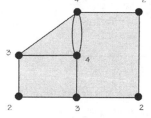

b)   This graph has no odd vertices and is connected, so it must have an Euler circuit.

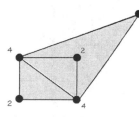

c)   This graph has even valences but the graph is disconnected, so it cannot have an Euler circuit.

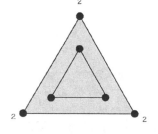

🖋 Question 2

Which of the following (if any) have an Euler circuit?

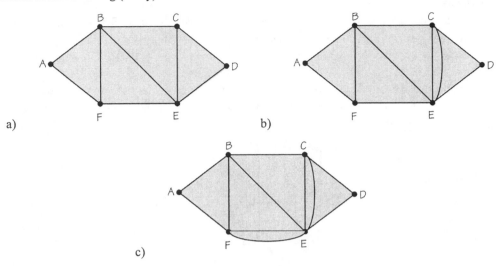

a)   b)

c)

Answer

c

🔑 Key idea

In finding an Euler circuit, never "disconnect" the graph by using an edge that is the only link between two parts of the graph not yet covered.

🔑 Key idea

Here, we are looking for an Euler circuit. Steps 1, 2, and 3 (*FBAC*) have been completed, and now we must decide where to proceed at vertex *C*. Proceeding to *E* "disconnects" *EF* from *CB*, *CD*, and *DB*. Proceeding to *B* is permissible. Proceeding to *D* is also permissible.

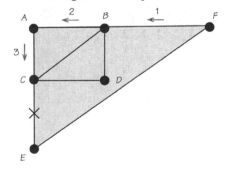

Section 1.3   Beyond Euler Circuits

⌥ Key idea

If a graph has odd vertices, then any circuit must reuse at least one edge. The **Chinese postman problem** involves finding a circuit that reuses as few edges as possible.

⌥ Key idea

This is a graph that contains odd vertices. One possible circuit follows the sequence of vertices *ADCABDEBDA*. This circuit reuses two edges: *AD* and *BD*. This is not the circuit that reuses the fewest edges for this graph.

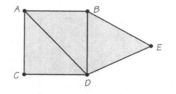

✍ Example D

Solve the Chinese postman problem for this graph—that is, find an Euler circuit of the graph that reuses the fewest edges.

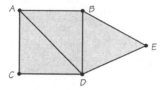

Solution

Since the graph does not have an Euler circuit, the best possible result is a circuit that reuses only one edge. One solution would be to start at *A*, then follow the sequence *ADBABEDCA*. This circuit reuses only one edge, *AB*.

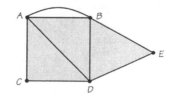

⌥ Key idea

Reusing an edge that joins two vertices is like adding a new edge between those vertices. Adding new edges for a circuit to produce an Euler circuit of a graph is called **eulerizing** the original graph.

Do You Know the Terms?

Cut out the following 14 flashcards to test yourself on Review Vocabulary. You can also find these flashcards at http://www.whfreeman.com/fapp7e.

Chapter 1 **Urban Services** **Chinese postman problem**	**Chapter 1** **Urban Services** **Circuit**
Chapter 1 **Urban Services** **Connected graph**	**Chapter 1** **Urban Services** **Digraph**
Chapter 1 **Urban Services** **Edge**	**Chapter 1** **Urban Services** **Euler circuit**
Chapter 1 **Urban Services** **Eulerizing**	**Chapter 1** **Urban Services** **Graph**

A path that starts and ends at the same vertex.	The problem of finding a circuit on a graph that covers every edge of the graph at least once and that has the shortest possible length.
A graph in which each edge has an arrow indicating the direction of the edge. Such directed edges are appropriate when the relationship is "one-sided" rather than symmetric (for instance, one-way streets as opposed to regular streets).	A graph is connected if it is possible to reach any vertex from any specified starting vertex by traversing edges.
A circuit that traverses each edge of a graph exactly once.	A link joining two vertices in a graph.
A mathematical structure in which points (called vertices) are used to represent things of interest and in which links (called edges) are used to connect vertices, denoting that the connected vertices have a certain relationship.	Adding new edges to a graph so as to make a graph that possesses an Euler circuit.

Chapter 1 Urban Services **Management science**	Chapter 1 Urban Services **Operations research**
Chapter 1 Urban Services **Optimal solution**	Chapter 1 Urban Services **Path**
Chapter 1 Urban Services **Valence (of a vertex)**	Chapter 1 Urban Services **Vertex**

Another name for management science.	A discipline in which mathematical methods are applied to management problems in pursuit of optimal solutions that cannot readily be obtained by common sense.
A connected sequence of edges in a graph.	When a problem has various solutions that can be ranked in preference order (perhaps according to some numerical measure of "goodness"), the optimal solution is the best-ranking solution.
A point in a graph where one or more edges end.	The number of edges touching that vertex.

Practice Quiz

1. What is the valence of vertex *B* in the graph below?

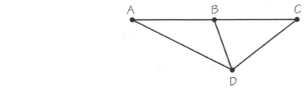

   a. 2

   b. 3

   c. 4

2. A graph is connected only if

   a. every vertex has an even valence.

   b. for every pair of vertices there is a path in the graph connecting these vertices.

   c. it has an Euler circuit.

3. For the graph below, which statement is correct?

   a. The graph has an Euler circuit.

   b. One new edge is required to eulerize this graph.

   c. Two new edges are required to eulerize this graph.

4. For which of the situations below is it most desirable to find an Euler circuit or an efficient eulerization of the graph?

   a. checking all the fire hydrants in a small town

   b. checking the pumps at the water treatment plant in a small town

   c. checking all the water mains in a small town

5. Consider the path represented by the sequence of numbered edges on the graph below. Which statement is correct?

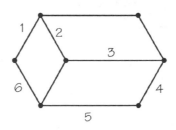

   a. The sequence of numbered edges forms an Euler circuit.

   b. The sequence of numbered edges traverses each edge exactly once, but is not an Euler circuit.

   c. The sequence of numbered edges forms a circuit, but not an Euler circuit.

6. What is the minimum number of duplicated edges needed to create a good eulerization for the graph below?

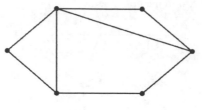

   a. 4 edges

   b. 3 edges

   c. 2 edges

7. Suppose the edges of a graph represent streets along which a postal worker must walk to deliver mail. Why would a route planner wish to find an Euler circuit or an efficient eulerization of this graph?

   a. to minimize the amount of excess walking the carrier needs to do

   b. to determine where postal drop boxes should be placed

   c. both a. and b.

8. If a graph has 10 vertices of odd valence, what is the absolute minimum number of edges that need to be added or duplicated to eulerize the graph?

   a. 5

   b. 10

   c. 0

9. Which option best completes the following analogy: A circuit is to a path as

   a. a vertex is to an edge.

   b. a digraph is to a graph.

   c. operations research is to management science.

10. Which of the following statements is true?
    I. If a graph is connected and has only even valences, then it has an Euler circuit.
    II. If a graph has an Euler circuit, then it must be connected and have only even valences.

   a. Only I is true.

   b. Only II is true.

   c. Both I and II are true.

Word Search

Refer to page 22 of your text to obtain the Review Vocabulary. There are 14 hidden vocabulary words/expressions in the word search below. All vocabulary words/expressions are represented separately. *Valence (of a vertex)* appears as *Valance*. It should be noted that spaces are removed.

```
F D N H I C F I F X K H H O Q E S Z M E E A W O D
T L O O D I I T D F O B L E T D B R A G T I E L I
C R R Z I I M Y I H K A R G R Q R F I S O H K D S
T R Z N G T R C I U E A Q R E A R L I S S F Y O O
S K L I R E U A P V C D S G M A E M W V S D R R L
E E A M A Q F L I G S R P E O A D E N S L U G X T
E R F C P S O A O P I R I I P Y S S N O E R O G P
Y T Z B H T I D D S C F I C E A F A M I L E P R J
A D F X P R M E H Y L S C T R E T A A A R W E A N
D F Z W A V A G G X A A G S H E T H N S E M R E Y
N Z I E R A N S Q D L I M T A O L W A W U G A T S
O N C X G L V I M H E Z L I T M X U G I L T T S A
O M R X M E L B O R P N A M T S O P E S E N I H C
O W G E H N L C A M Y A Y U E P K R M D R A O A H
S D E A S C Z M F C T I R B J E O O E I I A N C C
L D F Y S E E C Z E N G R G M O S T N F Z L S F E
H S N R R X W R A A O Q X F D O E G T N I J R G A
P T O E S S B D O G L T J S G E D I S E N D E N R
C A H S B P E M S T F L A P E A T S C C G P S B B
E N E E A M S I I R H P S H Y H C C I O V R E M H
S R E H N O H A R H Y F T E X E T R E V C J A I E
A A T S N P N D D Y R L L E N W C E N N X Y R J W
P H H L P Z S D D N W O S T H U N N C H N E C Q A
F A T I U E S C T R M P P K I Y A C E I N O H R T
H Q S S E C A T E I J G E T V D E H P L V Z C X J
```

1. _____

2. _____

3. _____

4. _____

5. _____

6. _____

7. _____

8. _____

9. _____

10. _____

11. _____

12. _____

13. _____

14. _____

Section 2.2   Traveling Salesman Problem

⌇ Key idea

The problem of finding this minimum-cost Hamiltonian circuit is called the **traveling salesman problem (TSP)**. It is a common goal in the practice of operations research.

Section 2.3   Helping Traveling Salesmen

⌇ Key idea

We use a variety of **heuristic** (or "fast") **algorithms** to find solutions to the TSP. Some are very good, even though they may not be optimal. Heuristic algorithms come close enough to giving optimal solutions to be important for practical use.

⌇ Key idea

The **nearest-neighbor algorithm** repeatedly selects the closest neighboring vertex not yet visited in the circuit (with a choice of edges, choose the one with the smallest weight), and returns to the initial vertex at the end of the tour.

⌇ Example F

Using the nearest-neighbor algorithm, finish finding a Hamiltonian circuit for this graph. We started at vertex C, proceeded to D, and then to B. When you've finished, calculate the total.

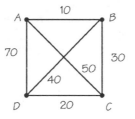

Solution

From B, the closest neighbor that has not yet been visited is A. All the vertices have thus been visited and you can return to starting point C. *CDBAC* is the complete circuit. The total of the weights of the edges in the path is 120.

✎ Question 3

Use the nearest-neighbor algorithm to find a Hamiltonian circuit for this graph starting at C. What is the total weight?

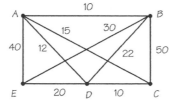

Answer

142

☞ Key idea

The **sorted-edges algorithm** (which, like nearest neighbor, is a **greedy algorithm**) is another heuristic algorithm that can lead to a solution that is close to optimal.

ᘓᔦ Example G

For this graph, what are the first three edges chosen according to the sorted-edges algorithm? Complete the circuit and calculate the total cost.

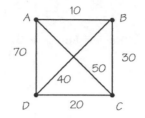

Solution

The edges *AB*, *CD*, and *BC* are the cheapest, and they do not close a loop or meet at a single vertex. This forces the choice of the expensive edge, *DA*, to return to the starting vertex. The complete circuit is *ABCDA*. The cost of all the edges in the tour adds up to 130.

✐ Question 4

Use the sorted-edges algorithm to find a Hamiltonian circuit for this graph. What is the total weight?

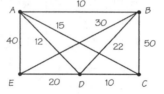

Answer
85

Section 2.4   Minimum-Cost Spanning Trees

☞ Key idea

A **tree** will consist of one piece and contains no circuits. A **spanning tree** is a tree that connects all vertices of a graph to each other with no redundancy (e.g., for a communications network.)

ᘓᔦ Example H

With a) as our original graph, which of the graphs shown in bold represent trees and/or spanning trees?

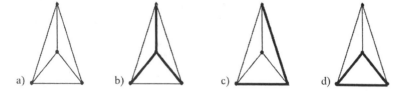

Solution

The graph in b) is a tree because it is a connected graph with no circuits, and it is a spanning tree because it includes all the vertices of the original graph; c) is a tree because it is a connected graph with no circuits, but is not a spanning tree because it does not include all the vertices of the original graph; and d) contains a circuit, so it cannot be a tree, and therefore cannot be a spanning tree.

⊶ Key idea

Two spanning trees for graph a) are shown in b) and c).

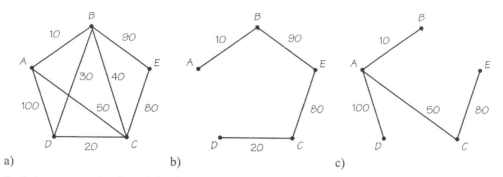

a)   b)   c)

Each is an example of a minimal **subgraph** connecting all the vertices in the original. In each case, removal of any edge will disconnect the graph.

⊶ Key idea

A **minimum-cost spanning tree** is most economical. **Kruskal's algorithm** produces one quickly. This algorithm adds links together in order of cheapest cost so that no circuits form and so that every vertex belongs to some link added.

⚭ Example I

Using this graph again, apply Kruskal's algorithm to find a minimum-cost spanning tree.

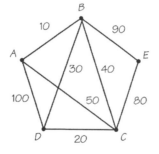

Solution

AB, *DC*, and *BD* are the cheapest edges and are chosen first. The next cheapest ones are *BC* and *AC*, but these would close loops. *EC* is the next in line, and that completes the tree.

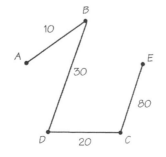

Homework Help

Exercises 1 – 9
Carefully read Section 2.1 before responding to these exercises. Remember that a Hamiltonian circuit visits each vertex once returning where it started.

Exercises 10 – 13
Carefully read Section 2.1 and Example B in this *guide* before answering these exercises.

Exercise 14
Carefully read the beginning of Section 2.1 before answering this exercise.

Exercises 15 – 16
Carefully read the definition of a Hamiltonian path in Exercise 15.

Exercise 17 – 19, 21
Review the difference between Euler circuits (Chapter 1) and Hamiltonian circuits (Section 2.1) before answering these exercises.

Exercise 20
Carefully read Section 2.1 before responding to this exercise. Graphs will vary.

Exercise 21
Carefully read the definition of a Hamiltonian path in Exercise 15.

Exercises 22 – 32
Carefully read Example 2 in Section 2.1 before responding to these exercises. These exercises involve applying the fundamental counting principle.

Exercises 33 – 67
Carefully read Section 2.3 and the examples before responding to these exercises.

Exercises 68 – 75
Carefully read Section 2.4 and the examples before responding to these exercises.

Do You Know the Terms?

Cut out the following 20 flashcards to test yourself on Review Vocabulary. You can also find these flashcards at http://www.whfreeman.com/fapp7e.

Chapter 2 Business Efficiency **Algorithm**	Chapter 2 Business Efficiency **Brute force method**
Chapter 2 Business Efficiency **Complete graph**	Chapter 2 Business Efficiency **Critical path**
Chapter 2 Business Efficiency **Fundamental principle of counting**	Chapter 2 Business Efficiency **Greedy algorithm**
Chapter 2 Business Efficiency **Hamiltonian circuit**	Chapter 2 Business Efficiency **Heuristic algorithm**

The method that solves the traveling salesman problem (TSP) by enumerating all the Hamiltonian circuits and then selecting the one with minimum cost.	A step-by-step description of how to solve a problem.
The longest path in an order-requirement digraph. The length of this path gives the earliest completion time for all the tasks making up the job consisting of the tasks in the digraph.	A graph in which every pair of vertices is joined by an edge.
An approach for solving an optimization problem, where at each stage of the algorithm the best (or cheapest) action is taken. Unfortunately, greedy algorithms do not always lead to optimal solutions.	A method for counting outcomes of multistage processes.
A method of solving an optimization problem that is "fast" but does not guarantee an optimal answer to the problem.	A circuit using distinct edges of a graph that starts and ends at a particular vertex of the graph and visits each vertex once and only once. A Hamiltonian circuit can start at any one of its vertices.

Chapter 2 Business Efficiency **Kruskal's algorithm**	Chapter 2 Business Efficiency **Method of trees**
Chapter 2 Business Efficiency **Minimum-cost Hamiltonian circuit**	Chapter 2 Business Efficiency **Minimum-cost spanning tree**
Chapter 2 Business Efficiency **Nearest-neighbor algorithm**	Chapter 2 Business Efficiency **NP-complete problems**
Chapter 2 Business Efficiency **Order-requirement digraph**	Chapter 2 Business Efficiency **Sorted-edges algorithm**

The problem of finding a minimum-cost Hamiltonian circuit in a complete graph where each edge has been assigned a cost (or weight).	A subgraph of a connected graph that is a tree and includes all the vertices of the original graph.
A number assigned to an edge of a graph that can be thought of as a cost, distance, or time associated with that edge.	A connected graph with no circuits.

Practice Quiz

1. Which of the following describes a Hamiltonian circuit for the graph below?

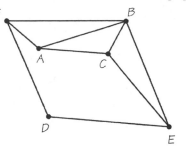

    a. *ABCEDCBFDA*

    b. *ABCDEFA*

    c. *ACBEDFA*

2. Using the nearest-neighbor algorithm and starting at vertex A, find the cost of the Hamiltonian circuit for the graph below.

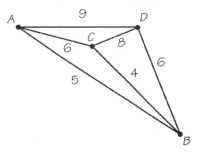

    a. 25

    b. 26

    c. Another answer

3. Using the sorted-edges algorithm, find the cost of the Hamiltonian circuit for the graph below.

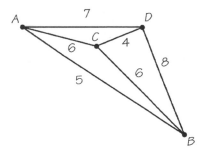

    a. 20

    b. 22

    c. 18

4. Using Kruskal's algorithm, find the minimum-cost spanning tree for the graph below. Which statement is true?

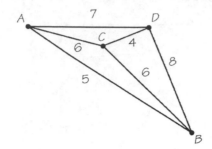

   a. Edges *AC* and *BD* are included.

   b. Edges *AC* and *AB* are included.

   c. Edges *CD* and *BD* are included.

5. Which of the following statements are true?
    I: It can be proved that Kruskal's algorithm always produces an optimal solution.
    II: If a graph has five vertices, its minimum-cost spanning tree will have four edges.

   a. Only I is true.

   b. Only II is true.

   c. Both I and II are true.

6. What is the critical path for the following order-requirement digraph?

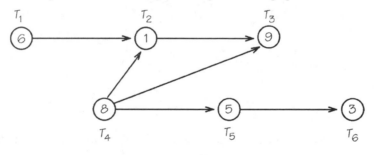

   a. T_4, T_2, T_3

   b. T_4, T_3

   c. T_1, T_2, T_3 and T_4, T_5, T_6 are both critical paths.

7. What is the earliest completion time (in minutes) for a job with the following order-requirement digraph?

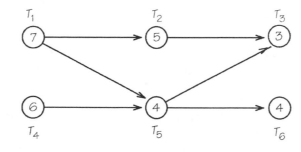

   a. 13 minutes

   b. 14 minutes

   c. 15 minutes

8. Which of the following statements are true?
    I: Trees never contain circuits.
    II: Trees are always connected and always include all the vertices of the larger graph.

   a. Only I is true.

   b. Only II is true.

   c. Neither I nor II is true.

9. Five small towns decide to set up an emergency communication system by connecting to each other with fiber optic cable. Which technique is most likely to be useful in helping them do this as cheaply as possible?

   a. finding an Euler circuit on a graph

   b. finding a Hamiltonian circuit on a graph

   c. finding a minimum-cost spanning tree on a graph

10. Scott Hochwald has three types of bread, four kinds of deli meat, and three types of cheese. How many different sandwiches could Scott Hochwald make?

   a. fewer than 15

   b. between 15 and 40

   c. more than 40

Word Search

Refer to pages 60 – 61 of your text to obtain the Review Vocabulary. There are 20 hidden vocabulary words/expressions in the word search below. All vocabulary appear separately. It should be noted that spaces are removed as well as apostrophes and hyphens. The abbreviation *TSP* does not appear in the word search.

```
U D N K K Z T S B U P F T I U C R I C N A I N O T L I M A H
L E E I U L M M H F E G A H Q R T E R S U M S T P A Z E E E
N P K G O G H G L R O A A M M S A R A W T I O Y O I Y O B R
T R G S A T T M G Z W S C R S F I L J F D N S B I S O U M I
R R D M O C I D L Q F T L V O G S A E T E I G A O O M O M S
A T G L S K R U S K A L S A L G O R I T H M R I Z L X O V L
V M N O K V O I E U T S R S G O P I D I S U T E O O A S M B
E H K E N M G R A S O R E Y K A O Q E K N M H F T R T E D P
L A N Y O E L I D F T O E D S S N E N P L C F M I E U C E A
I F M F T K A E E J E B D O H T E M E C R O F E T U R B S A
N V S E X C R I T I C A L P A T H F P C S S C P L Z N Z R V
G N I T N U O C F O E L P I C N I R P L A T N E M A D N U F
S P M Q Q R B R L E S E Z X A T S L Y S Z H H V D I I E H R
A R R N B E H P A R G E T E L P M O C F A A M T D S C N E T
L S G P T H G I E W E R E H F X S J O M A M Q S F S R I U M
E T R H H D I S E F I G F O G Q D E S F S I F T E K R E R N
S O R T E D E D G E S A L G O R I T H M R L I S N O V L Q Y
M E U E E T N L B J C S P T C B P E I E P T I A H B C E K J
A B I N E Q T R Z M C M H T I R O G L A O O E K O N M A W R
N C E B A F S X W H S E C P Y A G H I Y I N W S Z C F L F S
P L A G R E E D Y A L G O R I T H M E Z T I M Y L S H G F I
R L G H E U R I S T I C A L G O R I T H M A I C N S S O R M
O I X H R R A S E E R T F O D O H T E M E N V E I J S R O F
B O S R M R E E R T G N I N N A P S T S O C M U M I N I M C
L N U B E E N E B D Q Q F H N D L C J E S I O R S B X T N L
E R E I A Z F Y L Q Y P D Z C E D F E W S R B O P E E H E G
M D G L S I S M E L B O R P E T E L P M O C P N B M Z M L R
R X M U T E C H P A R G I D T N E M E R I U Q E R R E D R O
H I L E Y M J H N W O K S B K P P X L M F I T O H Y S C U E
N D F R P P M F E G A F O T T N G R E E R T G N I N N A P S
```

1. _____   11. _____

2. _____   12. _____

3. _____   13. _____

4. _____   14. _____

5. _____   15. _____

6. _____   16. _____

7. _____   17. _____

8. _____   18. _____

9. _____   19. _____

10. _____   20. _____

Chapter 3
Planning and Scheduling

Chapter Objectives

Check off these skills when you feel that you have mastered them.

- [ ] State the assumptions for the scheduling model.
- [ ] Compute the lower bound on the completion time for a list of independent tasks on a given number of processors.
- [ ] Describe the list-processing algorithm.
- [ ] Apply the list-processing algorithm to schedule independent tasks on identical processors.
- [ ] For a given list of independent tasks, compare the total task time using the list-processing algorithm for both the non-sorted list and also a decreasing-time list.
- [ ] When given an order-requirement digraph, apply the list-processing algorithm to schedule a list of tasks subject to the digraph.
- [ ] Explain how a bin-packing problem differs from a scheduling problem.
- [ ] Given an application, determine whether its solution is found by the list-processing algorithm or by one of the bin-packing algorithms.
- [ ] Discuss advantages and disadvantages of the next-fit bin-packing algorithm.
- [ ] Solve a bin-packing problem by the non-sorted next-fit algorithm.
- [ ] Solve a bin-packing problem by the decreasing-time next-fit algorithm.
- [ ] Discuss advantages and disadvantages of the first-fit bin-packing algorithm.
- [ ] Apply the non-sorted first-fit algorithm to a bin-packing problem.
- [ ] Apply the decreasing-time first-fit algorithm to a bin-packing problem.
- [ ] Discuss advantages and disadvantages of the worst-fit bin-packing algorithm.
- [ ] Find the solution to a bin-packing problem by the non-sorted best-fit algorithm.
- [ ] Find the solution to a bin-packing problem by the decreasing-time worst-fit algorithm.
- [ ] List two examples of bin-packing problems.
- [ ] Create a vertex coloring of a graph, and explain its meaning in terms of assigned resources.
- [ ] Find the chromatic number of a graph.
- [ ] Interpret a problem of allocation of resources with conflict as a graph, and find an efficient coloring of the graph.

41

Guided Reading

Introduction

We consider efficient ways to schedule a number of related tasks, with constraints on the number of workers, machines, space, or time available.

Section 3.1   Scheduling Tasks

⌐ Key idea

The **machine-scheduling problem** is to decide how a collection of tasks can be handled by a certain number of **processors** as quickly as possible. We have to respect both order requirements among the tasks and a priority list.

⌐ Key idea

The **list-processing algorithm** chooses a task for an available processor by running through the priority list in order, until it finds the **ready task**.

⌐ Example A

In this digraph, which tasks are ready

a) as you begin scheduling?

b) if just T_2 and T_3 have been completed?

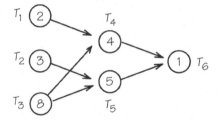

Solution

a)   At the start, only T_1, T_2, and T_3 have no required predecessors.

b)   With T_2 and T_3 completed, T_5 is ready but T_4 must wait for T_1. Also, T_6 must wait for T_4 and T_5.

⌐ Example B

Schedule the tasks in the digraph on two processors with priority list $T_1, T_2, T_3, T_4, T_5, T_6$, and determine the completion time.

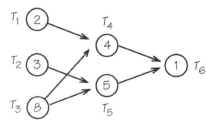

Continued on next page

Solution

T_1 and T_2 are first, then T_3. T_4 and T_5 must wait for completion of T_3.   Also, T_6 must wait for T_4 and T_5.   The completion time is 16.   The following is the schedule.

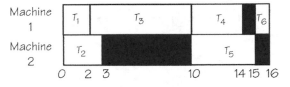

✐ Question 1

Schedule the tasks in the digraph on two processors with priority list T_1, T_2, T_3, T_4, T_5, T_6, T_7.   What is the completion time?

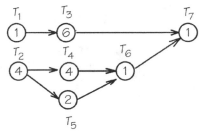

Answer

The completion time is 11.

⚷ Key idea

Different priority lists can lead to different schedules and completion times.

ᏮᏞ Example C

Schedule the tasks in the digraph on two processors with priority list T_6, T_5, T_4, T_3, T_2, T_1, and determine the completion time.

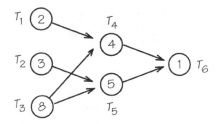

Solution

T_3 is the highest priority ready task and gets scheduled first, with T_2 next.   After T_2 is done, no task is ready except for T_1, so T_1 is scheduled.   When T_3 is done, T_6 is not ready so T_5 is scheduled, followed by T_4 and T_6.   The completion time is 14.   The following is the schedule:

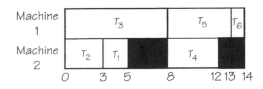

✐ Question 2

Schedule the tasks in the digraph on two processors with priority list $T_7, T_6, T_5, T_4, T_3, T_2, T_1$. What is the completion time?

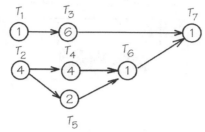

Answer

The completion time is 12.

🗝 Key idea

A schedule is optimal if it has the earliest possible total completion time. For example, a critical path in the order-requirement digraph may determine the earliest completion time.

⌇ Example D

Find a critical path in the digraph.

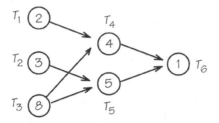

Is one of the schedules constructed optimal? If so, which one?

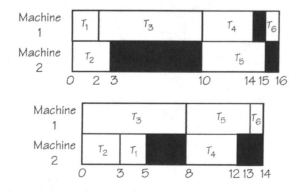

Solution

T_3, T_5, T_6 is the longest path in the digraph, and no scheduling can be completed in less time than the length of this path. Because the second schedule is completed at the same time of 14, it matches the critical path length of 14. Thus, the second schedule is optimal.

✏ Question 3

What is the length of the critical path?

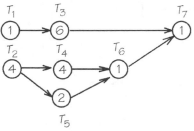

Answer

10

Section 3.2   Critical-Path Schedules

⌇ Key idea

If we can choose or change a priority list, then we have a chance to find an optimal schedule.

⌇ Key idea

The **critical-path scheduling** algorithm schedules first the tasks in a critical path.

𝒢 Example E

Use critical-path scheduling to construct a priority list for the tasks in this digraph.

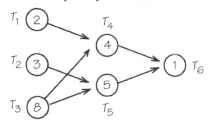

Solution

T_3 is at the head of a critical path. When you remove T_3 and its arrows, T_2 is the head of the remaining critical path T_2, T_5, T_6. Removing T_2 and its arrows makes T_1 the head of T_1, T_4, T_6. Finally, T_5 is the head of T_5, T_6. Thus the priority list is T_3, T_2, T_1, T_5, T_4, T_6.

𝒢 Example F

Construct the schedule for the tasks on two processors based on the critical path priority list from the above, T_3, T_2, T_1, T_5, T_4, T_6.

Solution

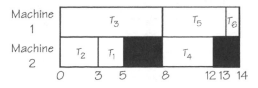

✎ Question 4

Use critical-path scheduling to construct a priority list for the tasks in this digraph. How much total time are the two machines not processing tasks?

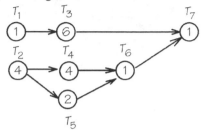

Answer

3

Section 3.3   Independent Tasks

☞ Key idea

When a set of tasks are **independent** (can be done in any order), we have a variety of available algorithms to choose a priority list leading to close-to-optimal scheduling. Some algorithms perform well in the **average-case**, but poorly in the **worst-case**.

☞ Key idea

The **decreasing-time-list algorithm** schedules the longest tasks earliest. By erasing the arrows in the digraph used throughout this chapter, we obtain a set of independent tasks.

ᏻ Example G

Construct a decreasing-time priority list. Use this list to schedule the tasks and determine the completion time on two machines.

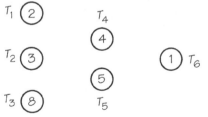

Solution

The list is T_3, T_5, T_4, T_2, T_1, T_6. Since we treat these as independent tasks, the task reference can be removed (keeping only the time). The schedule leads to a completion time of 12.

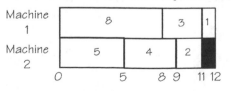

✐ Question 5

Construct a decreasing-time priority list. What is the completion time on two machines?

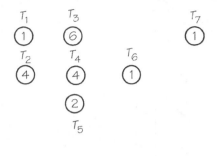

Answer

10

Section 3.4   Bin Packing

⊶ Key idea

With the **bin-packing problem**, we consider scheduling tasks within a fixed time limit, using as few processors as possible. This is like fitting boxes into bins of a certain size – but it is used in a variety of real-world applications.

⊶ Key idea

We have a variety of heuristic algorithms available to do the packing well if not optimally. Three important algorithms are **next fit (NF)**, **first fit (FF)**, and **worst fit (WF)**.

ᘯ Example H

Use the next-fit algorithm to pack boxes of sizes 4, 5, 1, 3, 4, 2, 3, 6, 3 into bins of capacity 8. How many bins are required?

Solution

There are six bins required.

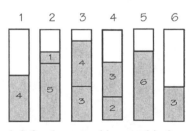

Bin 1 did not have enough space left for the second box, so bin 2 was used. There was enough room in bin 1 for the third box, but the NF heuristic doesn't permit us to go back to earlier bins. Once a bin is opened, it is used as long as the boxes fit – if they don't fit, a new bin is opened.

ᐉᔌ **Example I**

Now use the first-fit algorithm to pack boxes of sizes 4, 5, 1, 3, 4, 2, 3, 6, 3 into bins of capacity 8. How many bins are required?

Solution

There are five bins required.

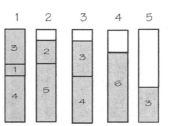

After opening bin 2 for the second box, we were able to go back to bin 1 for the next two boxes. The FF heuristic allows us to return to earlier bins while the NF does not.

⚷ **Key idea**

In the worst-fit (WF) algorithm we pack an item into a bin with the most room available. Although this algorithm can lead to the same number of bins as other algorithms, the items may be packed in a different order.

⚷ **Key idea**

WF is like FF in that it permits returning to earlier bins. However, in FF you always start back at the first bin and sequentially search for a bin that will accommodate this weight, while in WF you calculate the unused space in each available bin and select the bin with the maximum room.

✎ **Question 6**

Consider packing boxes sized 2, 6, 2, 6, 3, 4, 1, 4, 2 into bins of capacity 7. How many bins are required if we pack using

a) next-fit algorithm?

b) first-fit algorithm?

c) worst-fit algorithm?

Answer

a) 6

b) 5

c) 5

☞ Key idea

Each of these algorithms can be combined with **decreasing-time** heuristics, leading to the three algorithms **next-fit decreasing (NFD), first-fit decreasing (FFD),** and **worst-fit decreasing (WFD)**.

⌐ Example J

Use the first-fit decreasing algorithm to pack the boxes of sizes 4, 5, 1, 3, 4, 2, 3, 6, 3 into bins of capacity 8. How many bins are required?

Solution

First, rearrange the boxes in size decreasing order: 6, 5, 4, 4, 3, 3, 3, 2, 1.

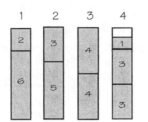

There are four bins required.

⌐ Example K

Is any of the algorithms you have used to pack the boxes sized 4, 5, 1, 3, 4, 2, 3, 6, 3 an optimal packing (that is, one using the fewest bins)?

Solution

In this case, FFD found the optimal packing. The amount of unused space is obviously less than the capacity of one bin. Therefore, no fewer than four bins could be used to hold all the boxes in this problem. However, neither FFD nor any of the other heuristics discussed in this section will necessarily find the optimal number of bins in an arbitrary problem.

✐ Question 7

Consider packing boxes sized 2, 6, 2, 6, 3, 4, 1, 4, 2 into bins of capacity 7. How many bins are required if we pack using

a)   next-fit decreasing algorithm?

b)   first-fit decreasing algorithm?

Answer

a)   5

b)   5

Section 3.5   Resolving Conflicts via Coloring

☞ Key idea

If we represent items to be scheduled (classes, interviews, etc) as vertices in a graph, then a **vertex coloring** of the graph can be used to assign resources, such as times or rooms, to the items in a conflict-free manner.

⌘ **Key idea**

The **chromatic number** of the graph determines the minimum amount of the resource that must be made available for a conflict-free schedule.

Here is a graph with five vertices, which is colored using four colors $\{A, B, C, D\}$.

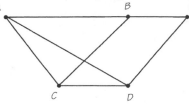

✍ **Example L**

Discuss the coloring of the following graph. Can you find a vertex coloring of the same graph with only three colors?

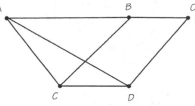

Solution

The vertex labeled A, for example, is connected to three other vertices, labeled C, B, D. No vertex is connected to four others. It is possible to find a vertex coloring of the same graph with only three colors. Here is one way to do it.

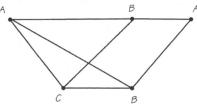

✏ **Question 8**

Suppose the following graph shows conflicts between animals A – H. If an edge connects two animals then they cannot be put in the same cage. Determine a suitable arrangement with a minimum number of cages. What is the minimum number of cages?

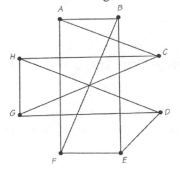

Answer

3

Homework Help

Exercises 1 – 3, 8
Answers will vary. Think of these as real-world situations.

Exercises 4 – 7, 9 – 26
Carefully read through Section 3.1 and 3.2 and their examples. Recall the length of the critical path is the longest path.

Exercises 27 – 41
Carefully read through Section 3.3 and its examples. With independent tasks, you do not need to be concerned about tasks preceding the ones you are scheduling.

Exercises 42 – 62
Carefully read through Section 3.4 and its examples. Be careful when applying the different methods of packing. Although the same number of bins may be used, they may be packed differently depending on the method.

Exercises 63 – 78
Carefully read through Section 3.5 and its examples. It is possible that two colorings of a graph are correct as long as they use the minimum number of colors.

Below are blocks that can be cut out in order to help you with scheduling machines and bin packing.

| 1 | 1 | 1 | 1 | 1 | 1 | 1 | 1 | 1 | 1 |

| 2 | 2 | 2 | 2 | 2 |

| 3 | 3 | 3 | 1 |

| 4 | 4 | 2 |

| 5 | 5 |

| 6 | 4 |

| 7 | 3 |

| 8 | 2 |

| 9 | 1 |

| 10 |

Do You Know the Terms?

Cut out the following 20 flashcards to test yourself on Review Vocabulary. You can also find these flashcards at http://www.whfreeman.com/fapp7e.

Chapter 3 Planning and Scheduling **Average-case analysis**	Chapter 3 Planning and Scheduling **Bin-packing problem**
Chapter 3 Planning and Scheduling **Chromatic number**	Chapter 3 Planning and Scheduling **Critical-path scheduling**
Chapter 3 Planning and Scheduling **Decreasing-time-list algorithm**	Chapter 3 Planning and Scheduling **First fit (FF)**
Chapter 3 Planning and Scheduling **First-fit decreasing (FFD)**	Chapter 3 Planning and Scheduling **Heuristic algorithm**

The problem of determining the minimum number of containers of capacity W into which objects of size $w_1, ..., w_n$ $(w_i \leq W)$ can be packed.	The study of the list-processing algorithm (more generally, any algorithm) from the point of view of how well it performs in all the types of problems it may be used for and seeing on average how well it does. *See also* worst-case analysis.
A heuristic algorithm for solving scheduling problems where the list-processing algorithm is applied to the priority list obtained by listing next in the priority list a task that heads a longest path in the order-requirement digraph. This task is then deleted from the order-requirement digraph, and the next task placed in the priority list is obtained by repeating the process.	The chromatic number of a graph *G* is the minimum number of colors (labels) needed in any vertex coloring of *G*.
A heuristic algorithm for bin packing in which the next weight to be packed is placed in the lowest-numbered bin already opened into which it will fit. If it fits in no open bin, a new bin is opened.	The heuristic algorithm that applies the list-processing algorithm to the priority list obtained by listing the tasks in decreasing order of their time length.
An algorithm that is fast to carry out but that doesn't necessarily give an optimal solution to an optimization problem.	A heuristic algorithm for bin packing where the first-fit algorithm is applied to the list of weights sorted so that they appear in decreasing order.

Chapter 3 Planning and Scheduling **Independent tasks**	Chapter 3 Planning and Scheduling **List-processing algorithm**
Chapter 3 Planning and Scheduling **Machine scheduling**	Chapter 3 Planning and Scheduling **Next fit (NF)**
Chapter 3 Planning and Scheduling **Next-fit decreasing (NFD)**	Chapter 3 Planning and Scheduling **Priority list**
Chapter 3 Planning and Scheduling **Processor**	Chapter 3 Planning and Scheduling **Ready task**

A heuristic algorithm for assigning tasks to processors: Assign the first ready task on the priority list that has not already been assigned to the lowest-numbered processor that is not working on a task.

Tasks are independent when there are no edges in the order-requirement digraph.

A heuristic algorithm for bin packing in which a new bin is opened if the weight to be packed next will not fit in the bin that is currently being filled; the current bin is then closed.

The problem of assigning tasks to processors so as to complete the tasks by the earliest time possible.

An ordering of the collection of tasks to be scheduled for the purpose of attaining a particular scheduling goal. One such goal is minimizing completion time when the list algorithm is applied.

A heuristic algorithm for bin packing where the next-fit algorithm is applied to the list of weights sorted so that they appear in decreasing order.

A task is called ready at a particular time if its predecessors, as given by the order-requirement digraph, have been completed by that time.

A person, machine, robot, operating room, or runway with time that must be scheduled.

Chapter 3
Planning and Scheduling

Vertex coloring

Chapter 3
Planning and Scheduling

Worst-case analysis

Chapter 3
Planning and Scheduling

Worst fit (WF)

Chapter 3
Planning and Scheduling

Worst-fit decreasing (WFD)

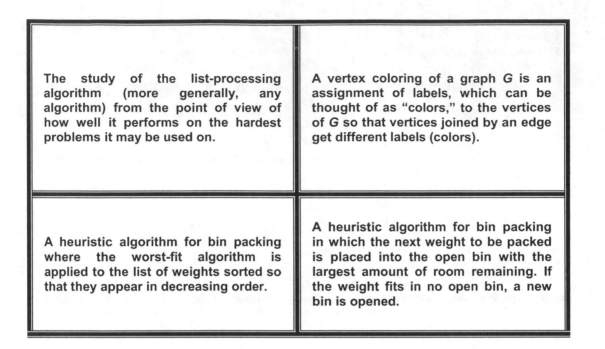

The study of the list-processing algorithm (more generally, any algorithm) from the point of view of how well it performs on the hardest problems it may be used on.

A vertex coloring of a graph G is an assignment of labels, which can be thought of as "colors," to the vertices of G so that vertices joined by an edge get different labels (colors).

A heuristic algorithm for bin packing where the worst-fit algorithm is applied to the list of weights sorted so that they appear in decreasing order.

A heuristic algorithm for bin packing in which the next weight to be packed is placed into the open bin with the largest amount of room remaining. If the weight fits in no open bin, a new bin is opened.

Practice Quiz

1. Given the order-requirement digraph below (with time given in minutes) and the priority list T_1, T_2, T_3, T_4, T_5, T_6, apply the list-processing algorithm to construct a schedule using two processors. How much time is required?

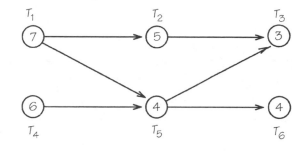

a. 13 minutes

b. 15 minutes

c. 16 minutes

2. Given the order-requirement digraph below (with time given in minutes) and the priority list T_1, T_2, T_3, T_4, T_5, T_6, apply the critical-path scheduling algorithm to construct a schedule using two processors. Which task is scheduled first?

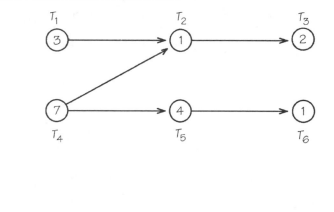

a. T_1

b. T_2

c. T_4

3. The director of a skating show has 25 skaters with varying length numbers to split into three segments, separated by intermissions. This job can be solved using:

a. the list-processing algorithm for independent tasks

b. the critical-path scheduling algorithm

c. the first-fit algorithm for bin packing

4. Use the decreasing-time-list algorithm to schedule the tasks below (times given in minutes) on 2 machines. How much time does the resulting schedule require?

Tasks: 5, 4, 7, 3, 8, 6, 2, 5, 8

a. 24 minutes

b. 25 minutes

c. more than 25 minutes

5. Use the first-fit algorithm to pack the following weights into bins that can hold no more than 10 pounds. How many bins are required?

    Weights: 5, 4, 7, 3, 8, 6, 2, 5, 8

    a. 5

    b. 6

    c. more than 6

6. Compare the results of the first-fit and the first-fit-decreasing algorithm to pack the following weights into bins that can hold no more than 10 pounds. Which statement is true?

    Weights: 5, 4, 7, 3, 8, 6, 2, 5, 8

    a. The two algorithms pack the items together in the same way.

    b. The two algorithms use the same number of bins, but group the items together in different ways.

    c. One algorithm uses fewer bins than the other.

7. Find the chromatic number of the graph below:

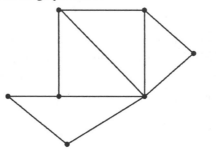

    a. 5

    b. 4

    c. 3

⌐ Key idea

The mixture chart can be translated into a set of inequalities (resource constraints and minimum constraints) along with a relation (often profit) to maximize.

✍ Example B

Write the resource constraints for cream and minimal constraints and the profit formula that applies to Dippy Dan's creamery shop.

Solution

Constraints: $x \geq 0$ and $y \geq 0$ (minimums); $\frac{1}{2}x + \frac{1}{4}y \leq 240$ (creme)

Profit formula: $P = \$0.75x + \$0.50y$

⌐ Key idea

In a mixture problem, we seek an optimal production policy for allocating limited resources to make a maximum profit. The resource constraints, together with the minimum constraints, can be used to draw a graph of the **feasible region**. A production policy is represented by a point in the feasible set, also called the feasible region.

⌐ Key idea

In determining a feasible region, you will need to graph a set of linear inequalities in the plane which involves the following:

- Graphing lines by the intercept method and determining which side to shade by using a test point (generally the origin);
- Graphing vertical and horizontal lines and determining which side to shade;
- Realizing that the minimum constraints of $x \geq 0$ and $y \geq 0$ imply that your graph is restricted to the upper right quadrant (quadrant I);
- Determining what region to finally shade, considering all inequalities.

✍ Example C

Draw the feasible region for Dippy Dan's creamery shop.

Solution

The minimum constraints $x \geq 0$ and $y \geq 0$ imply we are restricted to quadrant I.

We need to first graph $\frac{1}{2}x + \frac{1}{4}y \leq 240$.

The y-intercept of $\frac{1}{2}x + \frac{1}{4}y = 240$ can be found by substituting $x = 0$.

$$\tfrac{1}{2}(0) + \tfrac{1}{4}y = 240$$
$$0 + \tfrac{1}{4}y = 240$$
$$\tfrac{1}{4}y = 240 \Rightarrow y = 960$$

The y-intercept is $(0, 960)$.

The x-intercept of $\frac{1}{2}x + \frac{1}{4}y = 240$ can be found by substituting $y = 0$.

$$\tfrac{1}{2}x + \tfrac{1}{4}(0) = 240$$
$$\tfrac{1}{2}x + 0 = 240$$
$$\tfrac{1}{2}x = 240 \Rightarrow x = 480$$

The x-intercept is $(480, 0)$.

Continued on next page

We draw a line connecting these points. Testing the point $(0,0),$ we have the statement $\frac{1}{2}(0)+\frac{1}{4}(0)\le 240$ or $0\le 240$. This is a true statement, thus we shade the half-plane containing our test point, the down side of the line in quadrant I.

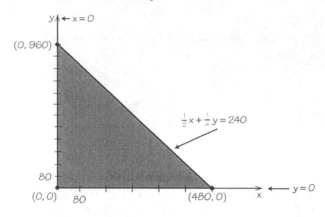

Section 4.2   Finding the Optimal Production Policy

⚬→ Key idea

According to the **corner point principle,** the optimal production policy is represented by a corner point of the feasible region. To determine the optimal production policy, we find the corner points of our region and evaluate the profit relation. The highest value obtained will indicate the optimal production policy − that is, how many of each product should be produced for a maximum profit.

⚭ Example D

Find the Dippy Dan's creamery shop optimal production policy.

Solution

We wish to maximize $\$0.75x + \$0.50y$.

Corner Point	Value of the Profit Formula: $\$0.75x + \$0.50y$
$(0,0)$	$\$0.75(0)$ + $\$0.50(0)$ = $\$0.00$ + $\$0.00$ = $\$0.00$
$(0,960)$	$\$0.75(0)$ + $\$0.50(960)$ = $\$0.00$ + $\$480.00$ = $\$480.00*$
$(480,0)$	$\$0.75(480)$ + $\$0.50(0)$ = $\$360.00$ + $\$0.00$ = $\$360.00$

Dan's optimal production policy: Make 0 containers of ice cream and 960 containers of sherbet for a profit of $480.

⚬→ Key idea

If one or more of the minimums are greater than zero, you will need to find a point of intersection between two lines. This can be done by **substitution**.

✍ Example E

Find the point of intersection between $x = 3$ and $3x + 2y = 171$.

Solution

By substituting $x = 3$ into $3x + 2y = 171$, we have the following.

$$3(3) + 2y = 171 \Rightarrow 9 + 2y = 171 \Rightarrow 2y = 162 \Rightarrow y = 81$$

Thus the point of intersection is $(3, 81)$.

✏ Question 1

a)   Find the point of intersection between $x = 20$ and $y = 30$.
b)   Find the point of intersection between $y = 5$ and $13x + 21y = 1678$.

Answer

a) $(20, 30)$
b) $(121, 5)$

✍ Example F

Dan has made agreements with his customers that obligate him to produce at least 100 containers of ice cream and 80 containers of sherbet. Find the Dippy Dan's creamery shop optimal production policy.

Solution

The minimums need to be adjusted in the mixture chart.

	Cream (240 pints)	Minimums	Profit
Ice cream, x containers	$\frac{1}{2}$	100	$0.75
Sherbet, y containers	$\frac{1}{4}$	80	$0.50

Constraints: $x \geq 100$ and $y \geq 80$ $(\text{minimums}); \frac{1}{2}x + \frac{1}{4}y \leq 240 (\text{cream})$

Profit formula: $P = \$0.75x + \$0.50y$

The point of intersection between $x = 100$ and $y = 80$ is $(100, 80)$.

The point of intersection between $x = 100$ and $\frac{1}{2}x + \frac{1}{4}y = 240$ can be found by substituting $x = 100$ into $\frac{1}{2}x + \frac{1}{4}y = 240$. We have $\frac{1}{2}(100) + \frac{1}{4}y = 240 \Rightarrow 50 + \frac{1}{4}y = 240 \Rightarrow \frac{1}{4}y = 190 \Rightarrow y = 760$. Thus, the point of intersection is $(100, 760)$.

The point of intersection between $y = 80$ and $\frac{1}{2}x + \frac{1}{4}y = 240$ can be found by substituting $y = 80$ into $\frac{1}{2}x + \frac{1}{4}y = 240$. We have $\frac{1}{2}x + \frac{1}{4}(80) = 240 \Rightarrow \frac{1}{2}x + 20 = 240 \Rightarrow \frac{1}{2}x = 220 \Rightarrow x = 440$. Thus, the point of intersection is $(440, 80)$.

Continued on next page

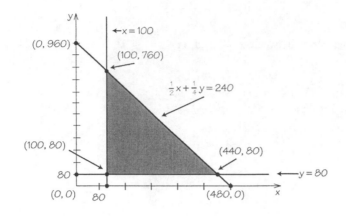

We wish to maximize $\$0.75x + \$0.50y$.

Corner Point	Value of the Profit Formula: $\$0.75x + \$0.50y$							
$(100, 80)$	$\$0.75(100)$	$+$	$\$0.50(80)$	$=$	$\$75.00$	$+$	$\$40.00$	$= \$115.00$
$(100, 760)$	$\$0.75(100)$	$+$	$\$0.50(760)$	$=$	$\$75.00$	$+$	$\$380.00$	$= \$455.00 *$
$(440, 80)$	$\$0.75(440)$	$+$	$\$0.50(80)$	$=$	$\$330.00$	$+$	$\$40.00$	$= \$370.00$

Dan's optimal production policy: Make 100 containers of ice cream and 760 containers of sherbet for a profit of $455.

↶ Example G

Suppose a competitor drives down Dan's price of sherbet to the point that Dan only makes a profit of $.25 per container. What is Dan's optimal production policy?

Solution

The feasible region and corner points will not change. However, we need to now maximize $\$0.75x + \$0.25y$.

Corner Point	Value of the Profit Formula: $\$0.75x + \$0.25y$							
$(100, 80)$	$\$0.75(100)$	$+$	$\$0.25(80)$	$=$	$\$75.00$	$+$	$\$20.00$	$= \$95.00$
$(100, 760)$	$\$0.75(100)$	$+$	$\$0.25(760)$	$=$	$\$75.00$	$+$	$\$190.00$	$= \$265.00$
$(440, 80)$	$\$0.75(440)$	$+$	$\$0.25(80)$	$=$	$\$330.00$	$+$	$\$20.00$	$= \$350.00 *$

Dan's optimal production policy: Make 440 containers of ice cream and 80 containers of sherbet for a profit of $350.

⚷ Key idea

With two resources to consider, there will be two resource constraints. The feasible region will generally be quadrilateral, with four corners to evaluate. To find the point of intersection between two lines (where neither are vertical or horizontal), one can use the **addition method**.

✍ Example H

Find the point of intersection between $5x + 2y = 17$ and $x + 3y = 6$.

Solution

We can find this by multiplying both sides of $x + 3y = 6$ by -5, and adding the result to $5x + 2y = 17$.

$$-5x - 15y = -30$$
$$\underline{5x + 2y = 17}$$
$$-13y = -13 \Rightarrow y = \tfrac{-13}{-13} = 1$$

Substitute $y = 1$ into $x + 3y = 6$ and solve to x. We have $x + 3(1) = 6 \Rightarrow x + 3 = 6 \Rightarrow x = 3$. Thus the point of intersection is therefore $(3,1)$.

✏ Question 2

Find the point of intersection between $7x + 3y = 43$ and $8x + 7y = 67$.

Answer

$(4,5)$

✍ Example I

Dan's creamery decides to produce raspberry versions of both its ice cream and sherbet lines. Dan is limited to, at most, 600 pounds of raspberries, and he adds one pound of raspberries to each container of ice cream or sherbet. Find the Dippy Dan's creamery shop optimal production policy. (Assume he still only makes a profit of $0.25 on a sherbet container and has non-zero minimums.)

Solution

The mixture chart is now as follows.

	Cream (240 pints)	Raspberries (600 lb)	Minimums	Profit
Ice cream, x containers	$\frac{1}{2}$	1	100	$0.75
Sherbet, y containers	$\frac{1}{4}$	1	80	$0.25

Constraints: $x \geq 100$ and $y \geq 80$ (minimums); $\frac{1}{2}x + \frac{1}{4}y \leq 240$ (cream); $x + y \leq 600$ (raspberries)

Profit formula: $P = \$0.75x + \$0.25y$

The y-intercept of $x + y = 600$ is $(0,600)$, and the x-intercept of $x + y = 600$ is $(600,0)$.

The point of intersection between $x = 100$ and $x + y = 600$ can be found by substituting $x = 100$ into $x + y = 600$. We have $100 + y = 600 \Rightarrow y = 500$. Thus, the point of intersection is $(100,500)$.

The final new corner point is the point of intersection between $x + y = 600$ and $\frac{1}{2}x + \frac{1}{4}y = 240$. We can find this by multiplying both sides of $\frac{1}{2}x + \frac{1}{4}y = 240$ by -4, and adding the result to $x + y = 600$.

$$-2x - y = -960$$
$$\underline{x + y = 600}$$
$$-x \quad\quad = -360 \Rightarrow x = 360$$

Substitute $x = 360$ into $x + y = 600$ and solve for y. We have, $360 + y = 600 \Rightarrow y = 240$. Thus, the point of intersection is $(360,240)$.

Continued on next page

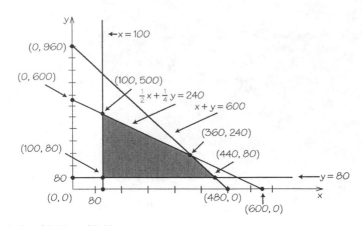

We wish to maximize $\$0.75x + \$0.50y$.

Corner Point	Value of the Profit Formula: $\$0.75x + \$0.25y$							
$(100, 500)$	$\$0.75(100)$	+	$\$0.25(500)$	=	$\$75.00$	+	$\$125.00$	=	$\$200.00$
$(360, 240)$	$\$0.75(360)$	+	$\$0.25(240)$	=	$\$270.00$	+	$\$60.00$	=	$\$330.00$
$(440, 80)$	$\$0.75(440)$	+	$\$0.25(80)$	=	$\$330.00$	+	$\$20.00$	=	$\$350.00\ast$
$(100, 80)$	$\$0.75(100)$	+	$\$0.25(80)$	=	$\$75.00$	+	$\$20.00$	=	$\$95.00$

Dan's optimal production policy: Make 440 containers of ice cream and 80 containers of sherbet for a profit of $350.

Note: With the additional constraint, there was no change because the optimal production policy already obeyed all constraints.

Section 4.3   Why the Corner Point Principle Works

⊶ Key idea

If you choose a theoretically possible profit value, the points of the feasible region yielding that level of profit lie along a profit line cutting through the region.

Profit Line: P = fixed value

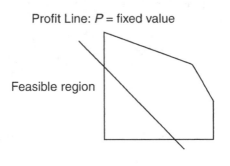

Feasible region

⌘ Key idea

Raising the profit value generally moves the profit line across the feasible region until it just touches at a corner, which will be the point with maximum profit, the optimal policy.

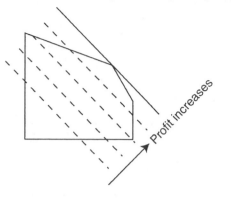

⌘ Key idea

This explains the corner point principle; it works because the feasible region has no "holes" or "dents" or missing points along its boundary.

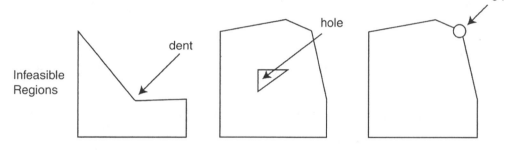

Section 4.4   Linear Programming: Life Is Complicated

⌘ Key idea

For realistic applications, the feasible region may have many variables ("products") and hundreds or thousands of corners. More sophisticated evaluation methods, such as the simplex method, must be used to find the optimal point.

⌘ Key idea

The **simplex method** is the oldest algorithm for solving linear programming. It was devised by the American mathematician George Dantzig in the 1940's. Years later, Narendra Karmarker devised an even more efficient algorithm.

∿ Example J

Dan's creamery decides to introduce a light ice cream (a third product) in his shop. Assume that a container of this light ice cream will require one-eighth pint of cream and sells at a profit of $1.00. There is no minimum on this new product, and it does not contain raspberries. Given all the constraints in Example I, determine the resource constraints and minimal constraints and the profit formula that applies to Dippy Dan's creamery shop.

Solution

Let x be the number of containers of ice cream, y be the number of containers of sherbet, and z be the number of containers of light ice cream.

	Cream (240 pints)	Raspberries (600 lb)	Minimums	Profit
Ice cream, x containers	$\frac{1}{2}$	1	100	$0.75
Sherbet, y containers	$\frac{1}{4}$	1	80	$0.25
Light ice cream, z containers	$\frac{1}{8}$	0	0	$1.00

Profit formula: $P = \$0.75x + \$0.25y + \$1.00z$

Constraints: $x \geq 100, \ y \geq 80, \text{ and } z \geq 0 \ (\text{minimums})$

$$\tfrac{1}{2}x + \tfrac{1}{4}y + \tfrac{1}{8}z \leq 240 \ (\text{cream})$$

$$x + y + 0z \leq 600 \, (\text{raspberries})$$

☛ Key idea

In order to solve applications with more than two products, you will need to access a program. Many of these are readily available on the Internet.

Section 4.5   A Transportation Problem: Delivering Perishables

☛ Key idea

A **transportation problem** involves supply, demand, and transportation costs. A supplier makes enough of a product to meet the demands of other companies. The supply must be delivered to the different companies, and the supplier wishes to minimize the shipping cost, while satisfying demand.

☛ Key idea

The amount of product available and the requirements are shown on the right side and the bottom of a table. These numbers are called **rim conditions**. A table showing costs (in the upper right-hand corner of a cell) and rim conditions form a **tableau**.

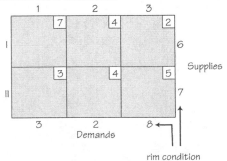

Each cell is indicated by its row and column. For example, the cell with a cost of 2 is cell (I, 3).

⊶ Key idea

The **northwest corner rule (NCR)** involves the following.

- Locate the cell in the far top left (initially that will be cell (I,1)).
- Cross out the row or column that has the smallest rim value for that cell.
- Place that rim value in that cell and reduce the other rim value by that smaller value.
- Continue that process until you get down to a single cell.
- Calculate the cost of this solution.

↶ Example K

Apply the Northwest Corner Rule to the following tableau and determine the cost associated with the solution.

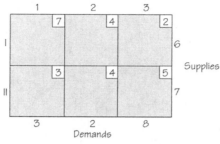

Solution

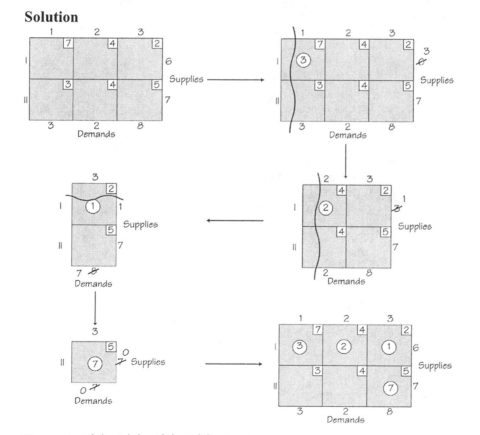

The cost is $3(7) + 2(4) + 1(2) + 7(5) = 21 + 8 + 2 + 35 = 66$.

✎ Question 3

Apply the Northwest Corner Rule to the following tableau to determine the cost associated with the solution.

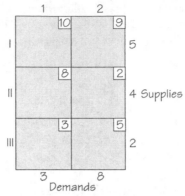

Answer

66

⚿ Key idea

The **indicator value of a cell** C (not currently a circled cell) is the cost change associated with increasing or decreasing the amount shipped in a circuit of cells starting at C. It is computed with alternating signs and the costs of the cells in the circuit.

✍ Example L

Determine the indicator values of the non-circled cells in Example K.

Solution

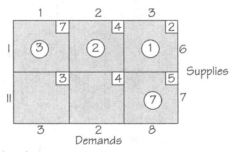

The indicator value for cell $(II, 2)$ is $4 - 5 + 2 - 4 = -3$.

The indicator value for cell $(II, 1)$ is $3 - 5 + 2 - 7 = -7$.

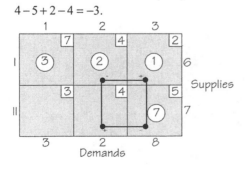

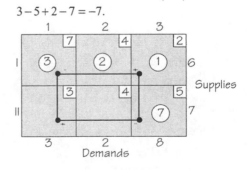

✐ Question 4

What is the indicator value of each of the non-circled cells in Question 3?

Answer

The indicator value for cell $(III,1)$ is -3 and for cell $(II,1)$ is 5.

☞ Key idea

If some indicator cells are positive and some are zero, there are multiple solutions for an optimal value.

Section 4.6   Improving on the Current Solution

☞ Key idea

The **stepping stone method** improves on some non-optimal feasible solution to a transportation problem. This is done by shipping additional amounts using a cell with a negative indicator value.

↩ Example M

Apply the stepping stone algorithm to determine an optimal solution for Example L. Consider both cells with negative indicator values, determine the new cost for each consideration and compare to the solution found using the Northwest Corner Rule.

Solution

Increasing the amount shipped through cell $(II,2)$ we have the following.

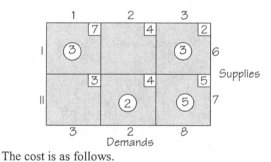

The cost is as follows.

$$3(7)+2(4)+5(5)+3(2)=21+8+25+6=60$$

Increasing the amount shipped through cell $(II,1)$ we have the following.

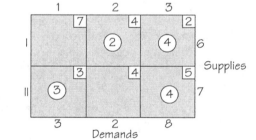

The cost is follows.

$$3(3)+2(4)+4(2)+4(5)=9+8+8+20=45.$$

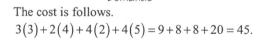

The cost can be reduced to 45 (compared to 66) when increasing the amount shipped through cell $(II,1)$.

✐ Question 5

Determine the cost involved when applying the stepping stone method to determine an optimal solution for Question 4 when compared to the cost found by the Northwest Corner Rule.

Answer

The cost can be reduced to 60 (compared to 66).

Homework Help

Exercise 1
In this exercise you will need to graph 6 lines. The first four are graphed by finding x- and y-intercepts and connecting these points. Recall, to find a y-intercept, substitute $x = 0$. To find an x-intercept, substitute $y = 0$. If a line is of the form $x = a$, then it is a vertical line (a is some number). If a line is of the form $y = b$, then it is a horizontal line (b is some number).

Exercises 2 – 3
In these exercises you will need to graph lines. You can find the point of intersection in Exercises 2 and 3(b) by the substitution method. In Exercise 3(a), you can use the addition method.

Exercises 4 – 5
In these exercises you will need to graph inequalities. First graph the line and choose a test point (usually the origin) to determine which side to shade. If your line is either vertical or horizontal, the direction of shading should be clear.

Exercises 6 – 8
In these exercises you will need to set up constraints (inequalities) given information.

Exercises 9 - 14
In these exercises you will need to graph a set of inequalities. Like in earlier exercises, you will need to first graph the line and choose a test point (usually the origin) to determine which side to shade. If your line is either vertical or horizontal, the direction of shading should be clear. The final graph should reflect the region common to all individual regions. Recall that the constraints of $x \geq 0$ and $y \geq 0$ indicate that you are restricted to the upper right quadrant created by the x-axis and y-axis.

Exercises 15 – 16
In these exercises you will need to determine if a point is in a feasible region or not. To do this, test the point in all of the inequalities. It must satisfy all of the constraints to be in the feasible region. If it does not satisfy any single constraint, it is not in the feasible region and you do not need to test any additional constraint.

Exercises 17 – 18
In these exercises you will need to evaluate each of the corner points in the profit relation. The point that yields that maximum value will determine the production policy. You should answer how many of each product should be made and what the maximum profit would be.
A table like the following may be helpful.

Corner Point	Value of the Profit Formula: $ x + $ y
(,)	\$ ( ) + \$ ( ) = \$ + \$ = \$
(,)	\$ ( ) + \$ ( ) = \$ + \$ = \$
(,)	\$ ( ) + \$ ( ) = \$ + \$ = \$

Optimal production policy: Make _____ skateboards and _____ dolls for a profit of \$_____.

Exercise 19
In this exercise you will need to graph two lines in the same plane. Like in earlier exercises, you will need to graph the lines by finding intercepts. To find the point of intersection, use the addition method.

Exercises 20 – 24

These exercises are essential to linear programming exercises that have two resource constraints. You will need to graph the two lines and find the points of intersection. To make the graphing process easier, find the x- and y- intercepts of both resource constraint lines. This will help you to determine a suitable scaling on the x- and y- axes. If properly drawn, you will readily see where the corner points are located. This will enable you to determine which two lines must be considered to find the corner points. In the case of the resource constraints, you will need to use the addition method to find the point of intersection. In Exercise 24, you can refer to the feasible regions in the referenced exercises or try each of the points in each constraint.

Exercises 25 – 28

These exercises take the previous type of exercises one step further. You will need to evaluate the corner points in the profit relation. The later part is like what was done in Exercises 17 – 18.

Exercise 29

In this exercise you will need to reply to each part. The goal is to consider realistic constraints where variables must be integer values.

Exercises 30 – 45

Using a simplex algorithm program will not be addressed in the solutions. If this approach is required by your instructor, ask for guidance as to where he or she expects that you obtain your solution (i.e. Internet or some other program supplied by him or her). These exercises address all the steps involved in determining the optimal solution. You must do the following.

- Define your variable;
- Set up a chart that shows the resource constraints, minimums, and profit;
- Define a set of inequalities and the profit relation that you wish to maximize;
- Draw the feasible region;
- Find all corner points (You will have two resource constraints in 36-41.);
- Evaluate the corner points in the profit relation to determine the optimal solution;
- State that solution clearly as to how many of each product should be made and what the maximum profit is.

Exercises 46 – 47

These are straightforward exercises. You are given a cost relation instead of a profit relation. You evaluate this cost relation with the same corner points of the referenced exercises. You will be seeking the minimum cost.

Exercises 48

If you think about where the x- and y- intercepts of $x + y = 0.5$ are located (non-integer values), then the answer should be apparent.

Exercises 49 – 51

Start by working these problems out as you did Exercises 30 – 45. After you've determined the optimal solution, see if the additional constraints cause the corner point that yielded the optimal solution to no longer be considered. If it is still considered, it will still yield the optimal solution. Otherwise, the process of evaluating corner points must be done again (with the new corner points). You will need to find indicator values in Exercises 53 – 54.

Exercises 52 – 54

In these exercises you will be applying the Northwest Corner Rule and finding the corresponding shipping costs. Carefully read the examples in Section 4.5. You may find the process easier if you "cancel" as you go from stage to stage as was done in Example K of this guide.

Exercises 55

Follow each step of this exercise. Recall from Chapter 3 the definition of a tree and from Chapter 1 the definition of a circuit.

Exercises 56
In this exercise you will need to find a solution using two methods and then compare the results. In part (a) follow the direction given for applying the minimum row entry method. You will determine the cost involved and compare to the solution found using the Northwest Corner Rule.

Exercises 57 – 58
In these exercises you will need to find a solution using the Northwest Corner Rule. You will next need to find the indicator value for each non-circled cell. This method is shown in Section 4.5 and in Example L of this guide. In Exercise 57, if the solution has a negative indicator in a cell, then the stepping stone method (Section 4.6) should be applied by shipping more via the cell that has the negative indicator value.

Do You Know the Terms?

Cut out the following 18 flashcards to test yourself on Review Vocabulary. You can also find these flashcards at http://www.whfreeman.com/fapp7e.

Chapter 4 **Linear Programming** ### Corner point principle	**Chapter 4** **Linear Programming** ### Feasible point
Chapter 4 **Linear Programming** ### Feasible region	**Chapter 4** **Linear Programming** ### Feasible set
Chapter 4 **Linear Programming** ### Indicator value of a cell	**Chapter 4** **Linear Programming** ### Linear programming
Chapter 4 **Linear Programming** ### Minimum constraint	**Chapter 4** **Linear Programming** ### Mixture chart

A possible solution (but not necessarily the best) to a linear programming problem. With just two products, we can think of a feasible point as a point on the plane.

The principle states that there is a corner point of the feasible region that yields the optimal solution.

Another term for feasible region.

The set of all feasible points, that is, possible solutions to a linear-programming problem. For problems with just two products, the feasible region is a part of the plane.

A set of organized methods of management science used to solve problems of finding optimal solutions, while at the same time respecting certain important constraints. The mathematical formulations of the constraints are linear equations and inequalities.

The change in cost due to shipping an increased or decreased amount using the cells in a transportation tableau that form a circuit consisting of circled cells together with a selected cell that is not circled. When an indicator value is negative, a cheaper solution can be found by shipping using this cell.

A table displaying the relevant data in a linear programming mixture problem. The table has a row for each product and a column for each resource, for any nonzero minimums, and for the profit.

An inequality in a mixture problem that gives a minimum quantity of a product. Negative quantities can never be produced.

Chapter 4
Linear Programming

Mixture problem

Chapter 4
Linear Programming

Northwest corner rule (NCR)

Chapter 4
Linear Programming

Optimal production policy

Chapter 4
Linear Programming

Profit line

Chapter 4
Linear Programming

Resource constraint

Chapter 4
Linear Programming

Rim conditions

A method for finding an initial but rarely optimal solution to a transportation problem starting from a tableau with rim conditions.   The amounts to be shipped between the suppliers and end users (demanders) are indicated by circling numbers in the cells in the tableau. The number of cells circled after applying the method will be equal to the number of rows plus the number of columns of the tableau minus 1.

A problem in which a variety of resources available in limited quantities can be combined in different ways to make different products. It is usually desired to find the way of combining the resources that produces the most profit.

In a two-dimensional, two-product, linear-programming problem, the set of all feasible points that yield the same profit.

A corner point of the feasible region where the profit formula has a maximum value.

The supply available (listed in a column at the right of a transportation tableau) and demands required (listed in a row at the bottom of a transportation tableau) in a transportation problem. The supply available are usually taken to exactly meet the demands required.

An inequality in a mixture problem that reflects the fact that no more of a resource can be used than what is available.

Chapter 4
Linear Programming

Simplex method

Chapter 4
Linear Programming

Stepping stone method

Chapter 4
Linear Programming

Tableau

Chapter 4
Linear Programming

Transportation Problem

A method for solving a transportation problem that improved the current solution, when it is not optimal, by increasing the amount shipped using a cell with a negative indicator value.

One of a number of algorithms for solving linear-programming problems.

A special type of linear programming problem where one has sources of supplies and users of, or demand for, these supplies. There is a cost to ship an item from a supplier to a user (demander). The goal is to minimize the total shipping cost to meet the demands from the supplies available.

A table for a transportation problem indicating the supplies available and demands required as well as the cost of shipping for a supplier to a user (demander). The amounts to be shipped from different suppliers to different users are indicated by circled cells in the tableau. The number of such circled cells is always the number of rows plus the number of columns diminished by one for the tableau.

Learning the Calculator

The graphing calculator can be used to graph a set of constraints and determine corner points.

Example

Use the graphing calculator to find the corner points for the following set of inequalities.

$$x \geq 0, \ y \geq 0, \ 2x+3y \leq 160, \ x+y \leq 60$$

Solution

Although the graphing calculator can locate points on the coordinate axes, because a proper window needs to be determined, it is easier to go ahead and calculate x- and y- intercepts of the lines.

In order to enter $2x+3y \leq 160$ and $x+y \leq 60$ into the calculator, we must solve each one for y.

$$2x+3y \leq 160 \Rightarrow 3y \leq 160-2x \Rightarrow y \leq (160-2x)/3$$

$$x+y \leq 60 \Rightarrow y \leq 60-x$$

After pressing ⌊Y=⌋, you can enter the equations.

```
Plot1  Plot2  Plot3
\Y1◘(160-2X)/3
\Y2◘60-X
\Y3=
\Y4=
\Y5=
\Y6=
\Y7=
```

To graph the first equation as an inequality, toggle over to the left of the line and press ⌊ENTER⌋ three times.

```
Plot1  Plot2  Plot3
▶Y1◘(160-2X)/3
\Y2◘60-X
\Y3=
\Y4=
\Y5=
\Y6=
\Y7=
```

Repeat for the other inequality.

```
Plot1  Plot2  Plot3
▶Y1◘(160-2X)/3
▶Y2◘60-X
\Y3=
\Y4=
\Y5=
\Y6=
\Y7=
```

By pressing ⌊WINDOW⌋ you will enter an appropriate window for Xmin, Xmax, Ymin and Ymax. Xmin and Ymin would both be 0 in most cases. Determine the x- and y-intercepts of the lines to assist you in determining a window as well as corner points.

$2x+3y=160$ has an x-intercept of $(80,0)$ and a y-intercept of $\left(0, 53\frac{1}{3}\right)$. $x+y=60$ has an x-intercept of $(60,0)$ and a y-intercept of $(0,60)$.

The following window is appropriate.

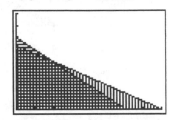

Next, we display the feasible region by pressing the GRAPH button.

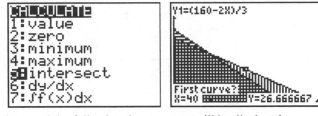

We know three of the four corner points. They are $(0,0)$, $(60,0)$, and $\left(0, 53\frac{1}{3}\right)$.   To find the fourth corner point, you will need to find the point of intersection between the lines $2x+3y=160$ and $x+y=60$.   To do this, you will need to press 2nd followed by TRACE. You then need to toggle down or press 5 followed by ENTER .

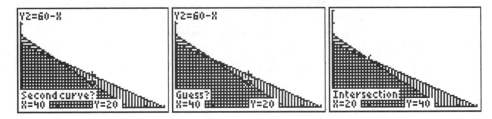

Press ENTER three times and the following three screens will be displayed.

The fourth corner point is therefore $(20, 40)$.

Practice Quiz

1. Where does the line $3x + 5y = 30$ cross the x-axis?

    a. at the point $(10,0)$

    b. at the point $(0,6)$

    c. at the point $(3,0)$

2. Where do the lines $2x + 3y = 11$ and $y = 1$ intersect?

    a. at the point $(1,3)$

    b. at the point $(4,1)$

    c. at the point $\left(0, \dfrac{11}{3}\right)$

3. Where do the lines $3x + 2y = 13$ and $4x + y = 14$?

    a. at the point $(2,3)$

    b. at the point $(1,10)$

    c. at the point $(3,2)$

4. Which of these points lie in the region $3x + 5y \le 30$?

    I. $(6,0)$

    II. $(1,2)$

    a. both I and II

    b. only II

    c. neither I nor II

5. What is the resource constraint for the following situation?
    Typing a letter (x) requires 4 minutes, and copying a memo (y) requires 3 minutes. A secretary has 15 minutes available.

    a. $x + y \le 15$

    b. $4x + 3y \le 15$

    c. $3x + 4y \le 15$

6. Graph the feasible region identified by the inequalities:

    $x \ge 0, \ y \ge 0, \ 3x + 2y \le 13, \ 4x + y \le 14$

    Which of these points is in the feasible region?

    a. $(3,3)$

    b. $(1,3)$

    c. $(5,0)$

7. What are the resource inequalities for the following situation?

    Producing a bench requires 2 boards and 10 screws. Producing a table requires 5 boards and 8 screws. Each bench yields $10 profit, and each table yields $12 profit. There are 25 boards and 60 screws available. x represents the number of benches and y represents the number of tables.

    a. $2x + 10y \leq 25$

    $5x + 8y \leq 60$

    $x \geq 0, y \geq 0$

    b. $2x + 10y \leq 10$

    $5x + 8y \leq 12$

    $x \geq 0, y \geq 0$

    c. $2x + 5y \leq 25$

    $10x + 8y \leq 60$

    $x \geq 0, y \geq 0$

8. What is the profit formula for the following situation?

    Producing a bench requires 2 boards and 10 screws. Producing a table requires 5 boards and 8 screws. Each bench yields $10 profit, and each table yields $12 profit. There are 25 boards and 60 screws available. x represents the number of benches and y represents the number of tables.

    a. $P = 25x + 60y$

    b. $P = \frac{1}{4}x + \frac{1}{5}y$

    c. $P = 10x + 12y$

9. Apply the Northwest Corner Rule to the following tableau and determine the cost associated with the solution.

    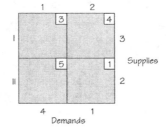

    a. cost: 28

    b. cost: 12

    c. cost: 15

10. Can the solution found by the Northwest Corner Rule in the last question be improved on?

    a. no.

    b. yes.

    c. not enough information.

Word Search

Refer to pages 163 – 164 of your text to obtain the Review Vocabulary. There are 18 hidden vocabulary words/expressions in the word search below. *Feasible set* and *Feasible region* will both appear. All vocabulary words/expressions appear separately. It should be noted that spaces are removed.

```
S G V N Y L R E L U R I J A J E I A N O L L U P V
A N F D M U E T O T L I R E H K C E A J G A N R H
C P H C O R N E R P O I N T P R I N C I P L E O T
R R R K L T N I A R T S N O C M U M I N I M N F R
S S M D H J A O V C A O T Q K E F E Q D D A N I E
I D P T E I D O H T E M E N O T S G N I P P E T S
A N T L V A E O G D E T H A V L F L M C K K Q F O
P A R F S T O W I E S K M Q U A E L B A T A P O U
E S O L E L U R R E N R O C T S E W H T R O N R R
N Q E S O M S N U L F E A S I B L E P O I N T M C
M E L B O R P N O I T A T R O P S N A R T S K U E
S M Q E E N T G R J S I D R S G L V K V R B N L C
B P F E S C M F P R O F I T L I N E F A G T R A O
O P T I M A L P R O D U C T I O N P O L I C Y S N
H E F A A C N B L A G V P Z C E N I O U H O E L S
J S E C F N E N O I G E R E L B I S A E F J N R T
S E A E S L G T F S K L E H W T M Z Q O I I H V R
O U S D E I I P A G D S A A E P P B N F A S P R A
T L I N E A R P R O G R A M M I N G L A X R R N I
A A B S D S X G O N O T M I X T U R E C H A R T N
E V L S Z E J M I X T U R E P R O B L E M R Q Q T
Q M E I X D I L M I E D O H T E M X E L P M I S T
I I S Y D A E I T O E O Y S S P R W A L X O S T R
S R E H O T P A M T W T A B I Y I D R F T E E N L
K N T N I E P E E Y C I L O P N O I T C U D O R P
```

1. _____   10. _____

2. _____   11. _____

3. _____   12. _____

4. _____   13. _____

5. _____   14. _____

6. _____   15. _____

7. _____   16. _____

8. _____   17. _____

9. _____   18. _____

Chapter 5
Exploring Data: Distributions

Chapter Objectives

Check off these skills when you feel that you have mastered them.

☐ Construct a histogram for a small data set.

☐ List and describe two types of distributions for a histogram.

☐ Identify from a histogram possible outliers of a data set.

☐ Construct a stemplot for a small data set.

☐ Calculate the mean of a set of data.

☐ Sort a set of data from smallest to largest and then determine its median.

☐ Determine the upper and lower quartiles for a data set.

☐ Calculate the five-number summary for a data set.

☐ Construct the diagram of a boxplot from the data set's five-number summary.

☐ Calculate the standard deviation of a small data set.

☐ Describe a normal curve.

☐ Given the mean and standard deviation of a normally distributed data set, compute the first and third quartiles.

☐ Explain the 68–95–99.7 rule.

☐ Sketch the graph of a normal curve given its mean and standard deviation.

☐ Given the mean and standard deviation of a normally distributed data set, compute the intervals in which the data set fall into a given percentage by applying the 68–95–99.7 rule.

Guided Reading

Introduction

Data, or numerical facts, are essential for making decisions in almost every area of our lives. But to use them for our purposes, huge collection of data must be organized and distilled into a few comprehensible summary numbers and visual images. This will clarify the results of our study and allow us to draw reasonable conclusions. The analysis and display of data are thus the groundwork for statistical inference.

⇥ Key idea

In a data set there are **individuals**. These individuals may be people, cars, cities, or anything to be examined.

⇥ Key idea

The characteristic of an individual is a **variable**. For different individuals, a variable can take on different values.

⌇ Example A

Identify the individuals and the variables in the following data set from a class roster.

Name	Age	Sex
Dan	16	Male
Edwin	17	Male
Adam	16	Male
Nadia	15	Female

Solution

The individuals are the names of the people on the class roster. The variables are their ages and sex.

⇥ Key idea

In this chapter, you will be doing **exploratory data analysis**. This combines numerical summaries with graphical display to see patterns in a set of data. The organizing principles of data analysis are as follows.

   1) Examine individual variables, and then look for relationships among variables.

   2) Draw a graph or graphs and add to it numerical summaries.

Section 5.1   Displaying Distributions: Histograms

⇥ Key idea

The **distribution** of a variable tells us what values the variable takes and how often it takes these values.

⇥ Key idea

The most common graph of a distribution with one numerical variable is called a **histogram**.

ᎧᏋ Example B

Construct a histogram given the following data. How many pieces of data are there?

Value	Count
5	2
10	5
15	7
20	3
25	1

Solution

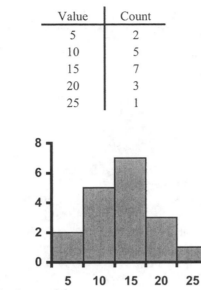

There are $2+5+7+3+1=18$ pieces of data.

ᎧᎢ Key idea

When constructing a histogram, each piece of data must fall into one **class**. Each class must be of equal width. For any given data set, there is more than one way to define the classes. Either you are instructed as to how to define the classes, or you must determine class based on some criteria.

ᎧᏋ Example C

Given the following exam scores, construct a histogram with classes of length 10 points.

40	50	50	53	55	55	55	58	60
60	63	65	68	70	70	73	75	75
78	78	83	85	85	88	90	95	96

Solution

It is helpful to first put the data into classes and count the individual pieces of data in each class. Since the smallest piece of data is 40, it makes sense to make the first class 40 to 49, inclusive.

Notice that the sum of the values in the count column should be 27 (total number of pieces of data). Also notice that some of the details of the scores are lost when raw data are placed in classes.

Class	Count
40 – 49	1
50 – 59	7
60 – 69	5
70 – 79	7
80 – 89	4
90 – 99	3

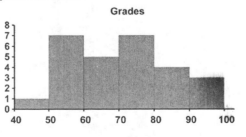

Section 5.2   Interpreting Histograms

⌥ Key idea

An important feature of a histogram is its overall **shape**. Although there are many shapes and overall patterns, a distribution may be **symmetric**, or it may be **skewed to the right** or **skewed to the left**.

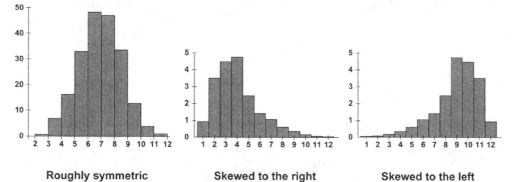

| Roughly symmetric | Skewed to the right | Skewed to the left |

If a distribution is skewed to the right, then the larger values extend out much further to the right. If a distribution is skewed to the left, then the smaller values extend out much further to the left. The easiest way to keep the two terms from being confused is to think of the direction of the "tail". If the tail points left, it is skewed to the left. If the tail points right, it is skewed to the right.

⌥ Key idea

Another way to describe a distribution is by its **center**. For now, we can think of the center of a distribution as the midpoint.

⌥ Key idea

Another way to describe a distribution is by its **spread**. The spread of a distribution is stating its smallest and largest values.

⌥ Key idea

In a distribution, we may also observe **outliers**; that is, a piece or pieces of data that fall outside the overall pattern. Often times determining an outlier is a matter of judgment. There are no hard and fast rules for determining outliers.

⌒ Example D

Given the following data regarding exam scores, construct a histogram. Describe its overall shape and identify any outliers.

Class	Count	Class	Count
0 – 9	1	50 – 59	6
10 – 19	0	60 – 69	8
20 – 29	0	70 – 79	7
30 – 39	0	80 – 89	5
40 – 49	3	90 – 99	2

Solution on next page

Solution

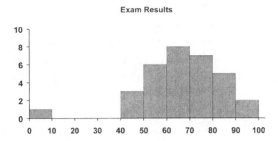

The shape is roughly symmetric. The score in the class 0 – 9, inclusive, is clearly an outlier. With a 0 on an exam, the most likely explanation is that the student missed the exam. It is also possible that the student was completely unprepared and performed poorly to obtain a very low score.

6⁄ Example E

Given the following data regarding exam scores, construct a histogram. Describe its overall shape and identify any outliers.

Class	Count	Class	Count
0 – 9	0	50 – 59	6
10 – 19	1	60 – 69	8
20 – 29	2	70 – 79	10
30 – 39	1	80 – 89	8
40 – 49	3	90 – 99	2

Solution

Exam Results

The shape is skewed to the left. There doesn't appear to be any outliers.

✎ Question 1

Given the following exam scores, describe the overall shape of the distribution and identify any outliers. In your solution, construct a histogram with class length of 5 points.

21	22	59	60	61	62	63	64	65
65	66	67	68	68	69	69	70	72
73	74	74	75	76	77	78	80	81
82	85	86	89	91	92	95		

Answer

The distribution appears to be skewed to the right. The scores of 21 and 22 appear to be outliers.

Section 5.3   Displaying Distributions: Stemplots

⊶ Key idea
A **stemplot** is a good way to represent data for small data sets. Stemplots are quicker to create than histograms and give more detailed information. Each value in the data set is represented as a stem and a leaf. The stem consists of all but the rightmost digit, and the leaf is the rightmost digit. A stemplot resembles a histogram turned sideways.

᎒◠ Example F
Given the following exam scores, construct a stemplot.

40	50	50	53	55	55	55	58	60
60	63	65	68	70	70	73	75	75
78	78	83	85	85	88	90	95	96

Solution
In the stemplot, the tens digit will be the stem and the ones digit will be the leaf.

```
4 | 0
5 | 0035558
6 | 00358
7 | 0035588
8 | 3558
9 | 056
```

✎ Question 2
The following are the percentages of salt concentrate taken from lab mixture samples. Describe the shape of the distribution and any possible outliers. This should be done by first rounding each piece of data to the nearest percent and then constructing a stemplot.

Sample	1	2	3	4	5	6	7
Percent	39.8	65.7	64.7	20.1	40.8	53.4	70.8
Sample	8	9	10	11	12	13	14
Percent	50.7	68.7	74.3	82.6	58.5	68.0	72.2

Answer
The distribution appears to be roughly symmetric with 20 as a possible outlier.

Section 5.4   Describing Center: Mean and Median

⊶ Key idea
The **mean** of a data set is obtained by adding the values of the observations in the data set and dividing by the number of data. If the observations are listed as values of a variable x (namely $x_1, x_2, ..., x_n$), then the mean is written as $\overline{x}$. The formula for the mean is $\overline{x} = \dfrac{x_1 + x_2 + ... + x_n}{n}$, where n represents the number of pieces of data.

ᎧᏗ Example G

Calculate the mean of each data set.

a)   123, 111, 105, 115, 112, 113, 117, 119, 114, 118, 111, 150, 147, 129, 138

b)   17, 15, 13, 2, 14, 15, 10, 1, 16, 16, 17, 22

Solution

a) $\bar{x} = \dfrac{123+111+105+115+112+113+117+119+114+118+111+150+147+129+138}{15}$

$\quad\quad = \dfrac{1822}{15} \approx 121.5$

b) $\bar{x} = \dfrac{17+15+13+2+14+15+10+1+16+16+17+22}{12} = \dfrac{158}{12} = 13.2$

✐ Question 3

Given the following stemplot, determine the mean. Round to the nearest tenth, if necessary.

```
1 | 259
2 | 3478
3 | 0334679
4 | 01259
5 | 46
6 | 1
7 | 3
```

Answer

$\bar{x} = 37$

⊶ Key idea

The **median**, *M*, of a distribution is a number in the middle of the data, so that half of the data are above the median, and the other half are below it. When determining the median, the data should be placed in order, typically smallest to largest. When there are n pieces of data, then the piece of data $\frac{n+1}{2}$ observations up from the bottom of the list is the median. This is fairly straightforward when n is odd. When there are n pieces of data and n is even, then you must find the average (add together and divide by two) of the two center pieces of data. The smaller of these two pieces of data is located $\frac{n}{2}$ observations up from the bottom of the list. The second, larger, of the two pieces of data is the next one in order or, $\frac{n}{2}+1$ observations up from the bottom of the list.

ᎧᏗ Example H

Determine the median of each data set below.

a)   123, 111, 105, 115, 112, 113, 117, 119, 114, 118, 111, 150, 147, 129, 138

b)   17, 15, 13, 2, 14, 15, 10, 1, 16, 16, 17, 22

Solution

For each of the data sets, the first step is to place the data in order from smallest to largest.

a)   105, 111, 111, 112, 113, 114, 115, **117**, 118, 119, 123, 129, 138, 147, 150

Since there are 15 pieces of data, the $\frac{15+1}{2} = \frac{16}{2} = 8^{\text{th}}$ piece of data, namely 117, is the median.

b)   1, 2, 10, 13, 14, **15, 15**, 16, 16, 17, 17, 22

Since there are 12 pieces of data, the mean of the $\frac{12}{2} = 6^{\text{th}}$ and 7th pieces of data will be the median. Thus, the median is $\frac{15+15}{2} = \frac{30}{2} = 15$. Notice, if you use the general formula $\frac{n+1}{2}$, you would be looking for a value $\frac{12+1}{2} = \frac{13}{2} = 6.5$ "observations" from the bottom. This would imply halfway between the actual 6th observation and the 7th observation.

✏ Question 4

Given the following stemplot, determine the median.

```
1 | 029
2 | 3478
3 | 03345679
4 | 012359
5 | 16
6 | 012
```

Answer

$M = 36.5$

Section 5.5   Describing Spread: The Quartiles

⌖ Key idea

The **quartiles** Q_1 (the point below which 25% of the observations lie) and Q_3 (the point below which 75% of the observations lie) give a better indication of the true spread of the data. More specifically, Q_1 is the median of the data to the left of M (the median of the data set). Q_3 is the median of the data to the right of M.

✑ Example I

Determine the quartiles Q_1 and Q_3 of each data set below.
a)   123, 111, 105, 115, 112, 113, 117, 119, 114, 118, 111, 150, 147, 129, 138
b)   17, 15, 13, 2, 14, 15, 10, 1, 16, 16, 17, 22

Solution

For each of the data sets, the first step is to place the data in order from smallest to largest.
a)   105, 111, 111, **112**, 113, 114, 115, 117, 118, 119, 123, **129**, 138, 147, 150

From Example H we know that the median is the 8th piece of data. Thus, there are 7 pieces of data below M. We therefore can determine Q_1 to be the $\frac{7+1}{2} = \frac{8}{2} = 4^{th}$ piece of data. Thus, $Q_1 = 112$. Now since there are 7 pieces of data above M, Q_3 will be the 4th piece of data to the right of M. Thus, $Q_3 = 129$.

b)   1, 2, **10, 13**, 14, 15, ↓ 15, 16, **16, 17**, 17, 22

From Example H we know that the median is between the 6th and 7th pieces of data. Thus, there are 6 pieces of data below M. Since $\frac{6+1}{2} = \frac{7}{2} = 3.5$, Q_1 will be the mean of 3rd and 4th pieces of data, namely $\frac{10+13}{2} = \frac{23}{2} = 11.5$. Now since there are 6 pieces of data above M, Q_3 will be the mean of the 3rd and 4th pieces of data to the right of M. Thus, $Q_3 = \frac{16+17}{2} = \frac{33}{2} = 16.5$.

✏ Question 5

Determine the quartiles Q_1 and Q_3 of each data set below.
a)   21, 16, 20, 6, 8, 9, 12, 15, 3, 15, 7, 8, 19
b)   14, 12, 11, 12, 24, 8, 6, 4, 8, 10

Answer

a) $Q_1 = 7.5$ and $Q_3 = 17.5$
b) $Q_1 = 8$ and $Q_3 = 12$

Section 5.6 The Five-Number Summary and Boxplots

⌐ Key idea
The **five-number summary** consists of the median (M), quartiles (Q_1 and Q_3), and extremes (high and low).

⌐ Key idea
A **boxplot** is a graphical (visual) representation of the five-number summary. A central box spans quartiles Q_1 and Q_3. A line in the middle of the central box marks the median, M. Two lines extend from the box to represent the extreme values.

✐ Example J
Given the following five-number summary, draw the boxplot.

$$200, 250, 300, 450, 700$$

Solution

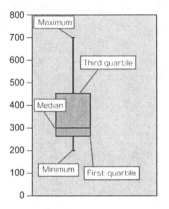

✐ Question 6
Given the following data, find the five-number summary and draw the boxplot.

$$12, 11, 52, 12, 15, 21, 17, 35, 16, 12$$

Answer

The five-number summary is 11, 12, 15.5, 21, 52.
The boxplot is as follows.

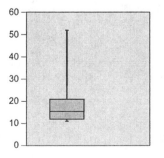

Section 5.7   Describing Spread: The Standard Deviation

⚡ Key idea

The **variance**, s^2, of a set of observations is an average of the squared differences between the individual observations and their mean value. In symbols, the variance of n observations

$(x_1, x_2, ..., x_n)$ is $s^2 = \dfrac{\left(x_1 - \overline{x}\right)^2 + \left(x_2 - \overline{x}\right)^2 + ... + \left(x_n - \overline{x}\right)^2}{n-1}$. Notice we divide by $n-1$.

⚡ Key idea

The **standard deviation**, s, of a set of observations is the square root of the variance and measures the spread of the data around the mean in the same units of measurement as the original data set. You should be instructed as to the method (spreadsheet, calculator with statistical capabilities, or by hand) required for calculating the variance and in turn the standard deviation.

⟲ Example K

Given the following data set, find the variance and standard deviation.

$$8.6, 7.2, 9.2, 5.6, 5.5, 4.4$$

Solution

Placing the data in order (not required, but helpful) we have the following hand calculations. Notice that $\overline{x} = \frac{40.5}{6} = 6.75$.

Observations x_i	Deviations $x_i - \overline{x}$	Squared deviations $\left(x_i - \overline{x}\right)^2$
4.4	$4.4 - 6.75 = -2.35$	$(-2.35)^2 = 5.5225$
5.5	$5.5 - 6.75 = -1.25$	$(-1.25)^2 = 1.5625$
5.6	$5.6 - 6.75 = -1.15$	$(-1.15)^2 = 1.3225$
7.2	$7.2 - 6.75 = 0.45$	$(0.45)^2 = 0.2025$
8.6	$8.6 - 6.75 = 1.85$	$(1.85)^2 = 3.4225$
9.2	$9.2 - 6.75 = 2.45$	$(2.45)^2 = 6.0025$
sum = 40.5	sum = 0.00	sum = 18.035

Thus, $s^2 = \dfrac{18.035}{6-1} = \dfrac{18.035}{5} = 3.607$ and $s = \sqrt{3.607} \approx 1.90$.

✎ Question 7

Given the following data set, find the variance and standard deviation.

$$3.41, 2.78, 5.26, 6.49, 7.61, 7.92, 8.21, 5.51$$

Answer

$s^2 \approx 4.169$ and $s = \sqrt{4.169} \approx 2.04$.

Section 5.8 Normal Distributions

⌗ Key idea

Sampling distributions, and many other types of probability distributions, approximate a bell curve in shape and symmetry. This kind of shape is called a normal curve, and can represent a **normal distribution**, in which the area of a section of the curve over an interval coincides with the proportion of all values in that interval. The area under any normal curve is 1.

⌗ Key idea

A normal curve is uniquely determined by its mean and standard deviation. The **mean** of a normal distribution is the center of the curve. The symbol μ will be used for the mean. The **standard deviation** of a normal distribution is the distance from the mean to the point on the curve where the curvature changes. The symbol σ will be use for the standard deviation.

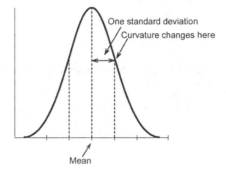

⌗ Key idea

The first quartile is located 0.67 standard deviation below the mean, and the third quartile is located 0.67 standard deviation above the mean. In other words, we have the following formulas.

$$Q_1 = \mu - 0.67\sigma \text{ and } Q_3 = \mu + 0.67\sigma$$

⌒ Example L

The scores on a marketing exam were normally distributed with a mean of 73 and a standard deviation of 12.

a) Find the third quartile (Q_3) for the test scores.

b) Find a range containing exactly half of the students' scores.

Solution

a) Since $Q_3 = \mu + 0.67\sigma = 73 + 0.67(12) = 73 + 8.04 = 81.04$, we would say the third quartile is 81.

b) Since 25% of the data lie below the first quartile and 25% of the data fall above the third quartile, 50% of the data would fall between the first and third quartiles. Thus, we must find the first quartile. Since $Q_1 = \mu - 0.67\sigma = 73 - 0.67(12) = 73 - 8.04 = 64.96$, we would say an interval would be $[65, 81]$.

Section 5.9   The 68 – 95 – 99.7 Rule

⚷ Key idea

The **68–95–99.7 rule** applies to a normal distribution. It is useful in determining the proportion of a population with values falling in certain ranges. For a normal curve, the following rules apply:
- The proportion of the population within one standard deviation of the mean is 68%.
- The proportion of the population within two standard deviations of the mean is 95%.
- The proportion of the population within three standard deviations of the mean is 99.7%.

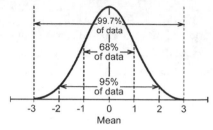

↷ Example M

The amount of coffee a certain dispenser fills 16 oz coffee cups with is normally distributed with a mean of 14.5 oz and a standard deviation of 0.4 oz.

a) Almost all (99.7%) cups dispensed fall within what range of ounces?

b) What percent of cups dispense less than 13.7 oz?

Solution

a)   Since 99.7% of all cups fall within 3 standard deviations of the mean, we find the following.

$$\mu \pm 3\sigma = 14.5 \pm 3(0.4) = 14.5 \pm 1.2$$

Thus, the range of ounces is 13.3 to 15.7.

b)   Make a sketch: 13.7 oz is two σ below μ; 95% are within 2σ of μ.

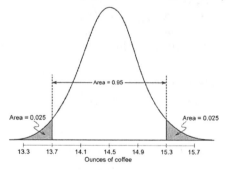

5% lie farther than 2σ. Thus, half of these, or 2.5%, lie below 13.7.

✐ Question 8

Look again at the marketing exam in which scores were normally distributed with a mean of 73 and a standard deviation of 12.

a)   Find a range containing 34% of the students' scores.

b)   What percentage of the exam scores were between 61 and 97?

Answer

a)   Either of the intervals [61, 73] or [73, 85]

b)   81.5%

Homework Help

Exercise 1
Carefully read the Introduction before responding to this exercise.

Exercises 2 – 3
Carefully read Section 5.2 before responding to these exercises. Pay special attention to the description of skewed distributions.

Exercise 4
Carefully read Sections 5.1 – 5.3 before responding to this exercise. First construct your classes and count individuals as described in Example 2 of your text. Include the outlier in your histogram. The following may be helpful in constructing your histogram. One possibility is to make the first class $6 \leq$ gas mileage < 11 or $11 \leq$ gas mileage < 16.

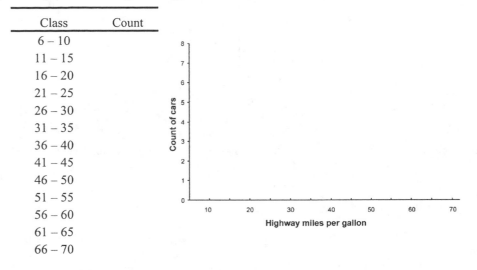

Class	Count
6 – 10	
11 – 15	
16 – 20	
21 – 25	
26 – 30	
31 – 35	
36 – 40	
41 – 45	
46 – 50	
51 – 55	
56 – 60	
61 – 65	
66 – 70	

Exercise 5
Carefully read Sections 5.1 – 5.2 before responding to this exercise. First construct your classes and count individuals as described in Example 2 of your text. Include the outliers in your histogram. The following may be helpful in constructing your histogram. One possibility is to make the first class $0.0 \leq$ emmissions < 2.0.

Class	Count
0.0 – 1.9	
2.0 – 3.9	
4.0 – 5.9	
6.0 – 7.9	
8.0 – 9.9	
10.0 – 11.9	
12.0 – 13.9	
14.0 – 15.9	
16.0 – 17.9	
18.0 – 19.9	

Pay special attention to the description of skewed distributions and outliers.

Exercise 6
Carefully read Section 5.2 before responding to this exercise. Pay special attention to the description of symmetric and skewed distributions. Think about how gender and right/left-handedness are distributed in real life.

Exercises 7 – 10
Carefully read Section 5.3 before responding to these exercises. Carefully read the description of how to describe each piece of data in Exercise 8. You may choose to use the following stems in the exercises.

Exercise 8		Exercise 9		Exercise 10	
0		10		48	
1		11		49	
2		12		50	
3		13		51	
		14		52	
		15		53	
		16		54	
		17		55	
		18		56	
		19		57	
		20		58	

Exercise 11
Carefully read Section 5.4 before responding to this exercise. Make sure to show all steps in your calculations, unless otherwise instructed.

Exercise 12
(a) Make the stemplot, with the outlier.

1	
2	
3	
4	
5	
6	
7	

(b) Calculate the mean. Use the stemplot to put the data in order from smallest to largest in order to find the median. Since there is an even number of pieces of data, you will need to examine two pieces of data to determine the median. Remove the outlier and recalculate the mean and determine the median of the 17 pieces of data. Compare the results with and without the outlier.

Exercise 13 – 14
Carefully read Section 5.2 before responding to these exercises. The following drawings may be helpful to show the relative locations of the median and the mean.

Skewed to the right   Skewed to the left

Exercises 15 – 16
Examples will vary.

Exercises 17 – 20
Carefully read Section 5.5 – 5.6 before responding to these exercises. Make sure to first put data in order from smallest to largest. Double check that you have accounted for all pieces of data. Pay special attention when you are dealing with an even number of pieces of data in determining the median. When determining quartiles, remember if there is an even number of pieces to the left of the mean, there will also be an even number of pieces to the right of the mean.

Exercise 21
Carefully read Section 5.5 – 5.6 before responding to this exercise. It would be helpful to create a stem plot to organize your data from smallest to largest.

```
 0 |
 1 |
 2 |
 3 |
 4 |
 5 |
 6 |
 7 |
 8 |
 9 |
10 |
11 |
12 |
13 |
14 |
15 |
16 |
17 |
18 |
19 |
```

Exercise 22
Carefully read Sections 5.5 – 5.6 before responding to this exercise. The data are already organized from smallest to largest.

Exercise 23
Look carefully at the referenced figure and compare as many features as possible.

Exercise 24
Your values of the median and quartiles may differ slightly from another student. Try rounding to the nearest thousand.

Exercise 25
Carefully read Sections 5.1 – 5.6 before responding to this exercise. Make sure to first put data in order from smallest to largest and round to the nearest whole number. Double check that you have accounted for all pieces of data. In part a, you can either create a histogram or a stemplot. Pay special attention when you are dealing with an even number of pieces of data in determining the median. When determining quartiles, remember if there is an even number of pieces to the left of the mean, there will also be an even number of pieces to the right of the mean.

Exercise 26
Approximate the bar heights. You will need to determine in which bar the 25%, 50%, and 75% marks occur for Q_1, M, and Q_3.

Exercises 27 – 28
Both of these exercises rely on the description of *interquartile range* given in Exercise 27.

Exercise 29

(a) Placing the data in order is helpful, but not required. If you are performing the calculations by hand, the following table may be helpful.

	Observations x_i		Deviations $x_i - \bar{x}$		Squared deviations $\left(x_i - \bar{x}\right)^2$
sum =		sum =	0.00	sum =	

(b) If your data are in order above, the median can easily be determined. If you have already worked Exercise 10, then the mean has already been calculated for comparison in this exercise.

Exercises 43 and 46
Carefully read Section 5.8 before responding to these exercises regarding quartiles of a normal distribution.

Exercise 47
Carefully read Sections 5.8 – 5.9 before responding to this exercise.

Exercise 50
Apply the formula given in part a for parts a and b and compare the results in part c.

Exercise 51
The following may be helpful in creating the stemplots.

Lengths of red flowers	Lengths of yellow flowers
37	34
38	35
39	36
40	37
41	38
42	
43	

Exercise 52
Arrange the data (separately) in order from smallest to largest in order to determine the five-number summary for each variety. Draw the boxplots and compare the skewness and the variabilities.

Exercises 53
Placing the data in order is helpful, but not required. When performing the calculations by hand, the following tables may be helpful. Note the order of red and yellow were switched for room considerations.

Yellow:

	Observations x_i		Deviations $x_i - \bar{x}$		Squared deviations $\left(x_i - \bar{x}\right)^2$
sum =		sum =		sum =	

Continued on next page

Exercises 53 continued
Red:

In the red data, you may choose to use 6 decimal place accuracy for $\bar{x}$ in order to calculate $x_i - \bar{x}$. You may also choose to round $\left(x_i - \bar{x}\right)^2$ to five decimal places. Round s^2 to four decimal places and s to three.

	Observations x_i		Deviations $x_i - \bar{x}$		Squared deviations $\left(x_i - \bar{x}\right)^2$
sum =		sum =		sum =	

Exercises 54 – 55
Carefully read Sections 5.8 – 5.9 before responding to these exercises.

Do You Know the Terms?

Cut out the following 19 flashcards to test yourself on Review Vocabulary. You can also find these flashcards at http://www.whfreeman.com/fapp7e.

Chapter 5 Exploring Data: Distributions **Boxplot**	Chapter 5 Exploring Data: Distributions **Distribution**
Chapter 5 Exploring Data: Distributions **Exploratory data analysis**	Chapter 5 Exploring Data: Distributions **Five-number summary**
Chapter 5 Exploring Data: Distributions **Histogram**	Chapter 5 Exploring Data: Distributions **Individuals**
Chapter 5 Exploring Data: Distributions **Mean**	Chapter 5 Exploring Data: Distributions **Median**
Chapter 5 Exploring Data: Distributions **Normal distributions**	Chapter 5 Exploring Data: Distributions **Outlier**

The pattern of outcomes of a variable. The distribution describes what values the variable takes and how often each value occurs.	A graph of the five-number summary. A box spans the quartiles, with an interior line marking the median. Lines extend out from this box to the extreme high and low observations.
A summary of a distribution that gives the median, the first and third quartiles, and the largest and smallest observations.	The practice of examining data for overall patterns and special features, without necessarily seeking answers to specific questions.
The people, animals, or things described by a data set.	A graph of the distribution of outcomes (often divided into classes) for a single variable. The height of each bar is the number of observations in the class of outcomes covered by the base of the bar. All classes should have the same width.
The midpoint of a set of observations. Half the observations fall below the median and half fall above.	The ordinary arithmetic average of a set of observations. To find the mean, add all the observations and divide the sum by the number of observations summed.
A data point that falls clearly outside the overall pattern of a set of data.	A family of distributions that describe how often a variable takes its values by areas under a curve. The normal curves are symmetric and bell-shaped. A specific normal curve is completely described by giving its mean and its standard deviation.

Chapter 5
Exploring Data: Distributions

Quartiles

Chapter 5
Exploring Data: Distributions

68 – 95 - 99.7 rule

Chapter 5
Exploring Data: Distributions

Skewed distribution

Chapter 5
Exploring Data: Distributions

Standard deviation

Chapter 5
Exploring Data: Distributions

Standard deviation of a normal curve

Chapter 5
Exploring Data: Distributions

Stemplot

Chapter 5
Exploring Data: Distributions

Symmetric distribution

Chapter 5
Exploring Data: Distributions

Variable

Chapter 5
Exploring Data: Distributions

Variance

In any normal distribution, 68% of the observations lie within 1 standard deviation on either side of the mean; 95% lie within 2 standard deviations of the mean; and 99.7% lie within 3 standard deviations of the mean.

The first quartile of a distribution is the point with 25% of the observations falling below it; the third quartile is the point with 75% below it.

A measure of the spread of a distribution about its mean as center. It is the square root of the average squared deviation of the observations from their mean.

A distribution in which observations on one side of the median extend notably farther from the median than do observations on the other side. In a right-skewed distribution, the larger observations extend farther to the right of the median than the smaller observations extend to the left.

A display of the distribution of a variable that attaches the final digits of the observations as leaves on stems made up of all but the final digit.

The standard deviation of a normal curve is the distance from the mean to the change-of-curvature points on either side.

Any characteristic of an individual.

A distribution with a histogram or stemplot in which the part to the left of the median is roughly a mirror image of the part to the right of the median.

A measure of the spread of a distribution about its mean. It is the average squared deviation of the observations from their mean. The square root of the variance is the standard deviation.

Learning the Calculator

Example 1

Construct a histogram given the following.

Value	Count
12	2
13	4
15	6
16	8
20	3

Solution

First enter the data by pressing the [STAT] button. The following screen will appear.

If there is data already stored, you may wish the clear it out. For example, if you wish to remove the data in L1, toggle to the top of the data and press [CLEAR] then [ENTER] . Repeat for any other data sets you wish to clear. Enter the new data being sure to press [ENTER] after each piece of data is displayed.

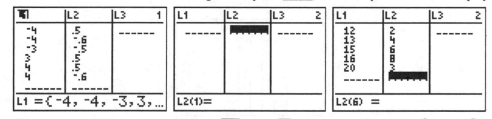

In order to display a histogram, you press [2nd] then [Y=] . This is equivalent to [STAT PLOT] . The following screen (or similar) will appear.

```
STAT PLOTS
1:Plot1...Off
   L L1    L2    □
2:Plot2...Off
   L L1    L2    □
3:Plot3...Off
   L L1    L2    □
4↓PlotsOff
```

You will need to turn a stat plot On and choose the histogram option (ᓕ). You will also need to make sure Xlist and Freq reference the correct data. In this case L1 and L2, respectively.

```
Plot1  Plot2  Plot3
On Off
Type: L L L  L
      L L  L
Xlist:L1
Freq:L2
```

Next, you will need to make sure that no other graphs appear on your histogram. Press $\boxed{Y=}$ and if another relation is present, either toggle to = and press enter to deselect or delete the relation.

Plot1 Plot2 Plot3	**Plot1** Plot2 Plot3	**Plot1** Plot2 Plot3
\Y₁■2X+1	\Y₁=2X+1	\Y₁=
\Y₂=	\Y₂=	\Y₂=
\Y₃=	\Y₃=	\Y₃=
\Y₄=	\Y₄=	\Y₄=
\Y₅=	\Y₅=	\Y₅=
\Y₆=	\Y₆=	\Y₆=
\Y₇=	\Y₇=	\Y₇=

You will next need to choose an appropriate window. By pressing $\boxed{\text{WINDOW}}$ you need to enter an appropriate window that includes your smallest and largest pieces of data. These values dictate your choices of Xmin and Xmax. Your choice of Xscl is determined by the kind of data you are given. In this case, the appropriate choice is 1. If you are given data such as 10, 12, 14, 16, and values such as 11, 13, and 15 are not considered then the appropriate choice would be 2 in order to make the vertical bars touch. In terms of choices for frequency, Ymin should be set at zero. Ymax should be at least as large as the highest frequency value. Your choice of Yscl is determined by how large the maximum frequency value is from your table.

```
WINDOW
 Xmin=10
 Xmax=22
 Xscl=1
 Ymin=0
 Ymax=10
 Yscl=1
 Xres=1
```

Next, we display the histogram by pressing the $\boxed{\text{GRAPH}}$ button.

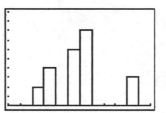

Notice that the histogram differs slightly from how a hand drawing should be. Ideally, the base of each rectangle should be shifted left by half of a unit.

Example 2

Given the following data, construct a histogram.

Class	Count
0 – 9	2
10 – 19	1
20 – 29	3
30 – 39	6
40 – 49	2

Solution

Follow the instructions in Example 1 in order to input data and set up the window to display the histogram. The width of the classes should be the Xscl in order to make the vertical bars touch. Also, in a case like this where you are given classes, use the left endpoint of the class as data pieces.

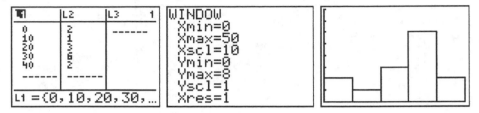

Example 3

Consider the following data.

$$21, 34, 55, 62, 54, 23, 34, 25, 50, 55, 52, 50$$

- Arrange the data in order from smallest to largest.
- Find the mean.

- Find the standard deviation.
- Find the five – number summary.
- Display the boxplot.

Enter the data, noting that there are 12 pieces of data. Make sure the location of the last entry corresponds to the total number of pieces of data.

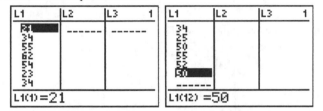

To arrange the data in order from smallest to largest, press the [STAT] button and choose the SortA(option which sorts the data in ascending order. Choose the appropriate data set (in this case L1) and then press [ENTER] . The calculator will display Done indicating the data is sorted.

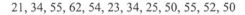

By pressing the [STAT] button, you can then view the data arranged in order by choosing the Edit option.

The data arranged from smallest to largest is as follows.

$$21, 23, 25, 34, 34, 50, 50, 52, 54, 55, 55, 62$$

To find the mean and standard deviation, press the [STAT] button. Toggle over to CALC and choose the 1-Var Stats option and then press the [ENTER]. You will get your home screen. Press [ENTER] again and you will then be able to determine the mean and standard deviation.

```
EDIT CALC TESTS
1:1-Var Stats
2:2-Var Stats
3:Med-Med
4:LinReg(ax+b)
5:QuadReg
6:CubicReg
7↓QuartReg
```

```
SortA(L₁)
              Done
1-Var Stats
```

```
1-Var Stats
x̄=42.91666667
Σx=515
Σx²=24421
Sx=14.51931837
σx=13.901189
↓n=12
```

The mean is   (approximately 42.917) and the standard deviation is Sx (approximately 14.519).

To determine the five – number summary, from the last screen press the down arrow ([▾]) five times.

```
1-Var Stats
↑n=12
 minX=21
 Q₁=29.5
 Med=50
 Q₃=54.5
 maxX=62
```

The five – number summary is 21, 29.5, 50, 54.5, 62.

To display the box plot, press [2nd] then [Y=]. This is equivalent to [STAT PLOT]. You will need to choose ⊡ for boxplot. Make sure the proper data are chosen for Xlist and Freq should be set at 1.

Choose an appropriate window for Xmin and Xmax based on the minimum and maximum values. The values you choose for Ymin and Ymax do not have an effect on the boxplot. You may choose values for Xscl and Yscl based on appearance of the axes. Display boxplot by pressing the [GRAPH] button.

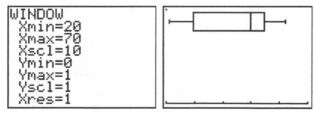

Practice Quiz

1. The weights (in pounds) of your cousins are: 120, 89, 108, 76, 21. Which are the outliers?

    a. 21 only.

    b. 120 only

    c. both 120 and 21

2. Below is a stemplot of the ages of adults on your block. Which statement is true?

    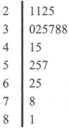

2	1125
3	025788
4	15
5	257
6	25
7	8
8	1

    a. The stemplot is roughly symmetric.
    b. The stemplot is skewed to the higher ages.
    c. The two oldest people are outliers.

3. Here are 7 measured lengths (in inches): 13, 8, 5, 3, 8, 9, 12. Find their median.
    a. 3
    b. 8
    c. 8.3

4. Here are 7 measured lengths (in inches): 13, 8, 5, 3, 8, 9, 12. Find their mean.
    a. 3
    b. 8
    c. 8.3

5. The boxplot graph always includes the
    a. mean and median.
    b. quartiles and the standard deviation.
    c. quartiles and the median.

6. The percentage of scores on a standardized exam that lie between the first and third quartiles is
    a. 25%.
    b. 50%.
    c. 75%.

7. If the mean of the data 2, 4, 6, 3, 5, 8, 7 is 5, what is its standard deviation?
    a. $\sqrt{\frac{12}{7}}$
    b. 4
    c. 2

8. The scores on a marketing exam were normally distributed with a mean of 67 and a standard deviation of 9. Find the first quartile (Q_1) for the test scores.

   a. 58

   b. 61

   c. 25

9. Given the following data, find the five-number summary.

   $$5, 8, 12, 15, 11, 21, 9, 12$$

   a. 5, 8.5, 11.5, 13.5, 21

   b. 5, 8.5, 12, 13, 21

   c. 5, 12, 13, 21, 12

10. The amount of coffee a certain dispenser fills 12 oz coffee cups with is normally distributed with a mean of 10.9 oz and a standard deviation of 0.2 oz. What percent of cups dispense more than 11.1 oz?

    a. 68%

    b. 5%

    c. 16%

Word Search

Refer to pages 207 – 208 of your text to obtain the Review Vocabulary. There are 17 hidden vocabulary words/expressions in the word search below. *Standard deviation of a normal curve* and *68-95-99.7 rule* were both omitted from the word search. It should be noted that spaces and hyphens are removed.

```
I Y E Q T E L L I A T S O P M S T S Y V G N I I A S
M U E R E R F J U O R M T E I T Z C N F P E N G I A
G R G E N I P Q G V O E I J R P S O G Q S E I S N K
S R J R D P T F O E I A K N E H A T E T P H I K B T
I O H I S T O G R A M T C O Z F L I A L C A M E L H
X A S C S J Y O E L B A I R A V C N I K R Q H W L D
S J Y F T O Z H T X S P G M W O D H L O I A P E C E
O N M A E U D R G F L N T A N A T O L P X O B D U G
E I M I M T E G R S E I K L R E A M O L J I S D A K
E O E X P L O R A T O R Y D A T A A N A L Y S I S E
T F T U L I A M M S G K D I P E Q V S D H M J S S F
M T R O O E L E E S E Y S T H E H L I E E T T X P
X Q I I T R T O D L V Z D T E C N E A S A A Q R G E
A D C I I S Z N I I D E N R M E S A U T O N H I F O
S F D D T X N T A T A T F I L C F R D R E E A B E F
Y I I E W K V T N R D F R B A E S T I I L C A U E N
O T S R K G I Z X A H E K U F H R O V B G N A T A J
Y E T O T O Z L E U Z C A T S I X T I U V A A I D I
E M R V N N I F X Q M I O I O U R H D T E I E O A E
I Y I E N L A P T E K R U O N S O L N I O R Z N N S
V D B S N E E A Q F S O E N M H C X I O J A C H H W
L O U A D N M Y R A M M U S R E B M U N E V I F J I
A I T Y L F O S T D B I O O S C M A F N N F U I G S
K S I P N K D S S Q G D S W H O U G N A R F R O M X
M N O M U A Z S I O E Y Q K N T Q S O H O H S L M F
P X N S A I Q P U R S R R D R Z E Q S A E K R R R H
```

1. _____

2. _____

3. _____

4. _____

5. _____

6. _____

7. _____

8. _____

9. _____

10. _____

11. _____

12. _____

13. _____

14. _____

15. _____

16. _____

17. _____

Chapter 6
Exploring Data: Relationships

Chapter Objectives

Check off these skills when you feel that you have mastered them.

☐ Draw a scatterplot for a small data set consisting of pairs of numbers.

☐ From a scatterplot, draw an estimated line of best fit.

☐ Describe how the concept of distance is used in determining a least-squares regression line.

☐ Use the given equation of a regression line to predict response (y) values from given explanatory (x) values.

☐ Calculate the correlation between two quantitative variables, one explanatory and one response, from a data set.

☐ Understand the significance of the correlation between two variables, and estimate it from a scatterplot.

☐ Understand correlation and regression describe relationships that need further interpretation because association does not imply causation and outliers have an effect on these relationships.

Guided Reading

Introduction

Relationships between variables exist in almost every area of our lives. For example, insurance companies use relationships between variables to determine appropriate annual rates. The medical community uses relationships between variables to help project the effects of drugs, certain foods, and exercise on certain aspects of our lives such as lifespan. By determining a relationship between variables and the strength of that relationship, one can draw reasonable conclusions.

⊶ Key idea
We will be using data sets that have two types of variables. A **response variable** measures an outcome or result of a study. An **explanatory variable** is a variable that we think explains or causes changes in the response variable. Typically we think of the explanatory variable as x and the response variable as y.

Section 6.1   Displaying Relationships: Scatterplots

⌐ Key idea

Graphs are useful for recognizing connections between two variables. A **scatterplot** is the simplest such representation, showing the relationship between an explanatory variable (on the horizontal axis) and a response variable (on the vertical axis).

⌐ Key idea

We look for an **overall pattern** in the scatterplot. The pattern can be described by the following.
- *form*: straight – line, for example
- *direction*: positive association or negative association (slope of a line)
- *strength*: A stronger relationship would yield points quite close to the line, a weaker one would have more points scattered around the line.

⌐ Key idea

We also look for striking **deviations** from the pattern in the scatterplot. An important kind of deviation is an *outlier*, an individual value that falls outside the overall pattern of the relationship.

⌐ Example A

Draw a scatterplot showing the relationship between the observed variables x and y, with the data given in the table below. Make some observations of the overall pattern in the scatterplot.

x	2	4	1	5	7	9	8
y	2	5	2	7	4	8	6

Solution

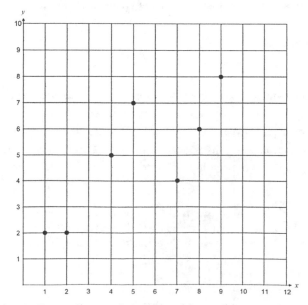

It appears that the points indicate a linear relationship with a positive association.

Section 6.2   Regression Lines

✑ Key idea

A straight line drawn through the heart of the data and representing a trend is called a **regression line**, and can be used to predict values of the response variable.

✑ Key idea

The equation of a regression line will be $y = a + bx$, where a is the y-intercept and b is the slope of the line.

ᏽ Example B

Starting with the scatterplot of the data from the previous section, draw the regression line $y = 1.66 + 0.622x$ through the data (obtained from a computer program). Use the graph to predict the value of y if the x-value is 11.

Solution

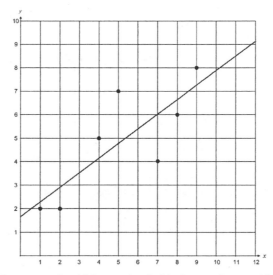

According to the graph, it appears that if the x-value is 11, the y-value would be approximately 8.5. Using the equation of the regression line we have the following.

$$y = 1.66 + 0.622(11) = 1.66 + 6.842 = 8.502$$

✐ Question 1

Given the following data with regression line $y = 8.19 - 0.477x$ (obtained from a computer program). Determine which point is closest to the regression line and which point is farthest. Do this by making a scatterplot, drawing the regression line, and visually determining which point is closest and which point is farthest from the line.

x	10	1	6	7	4	9	3	5
y	5	8	4	3	6	4	8	6

Answer

$(7,3)$ appears to be farthest away and $(9,4)$ appears to be closest to the regression line.

Section 6.3 Correlation

☞ Key idea

The correlation, r, measures the strength of the linear relationship between two quantitative variables; r always lies between -1 and 1.

☞ Key idea

Positive r means the quantities tend to increase or decrease together; **negative r** means they tend to change in opposite directions, one going up while the other goes down. If **r is close to 0**, that means the quantities are fairly independent of each other.

⌇ Example C

Give a rough estimate of the correlation between the variables in each of these scatterplots:

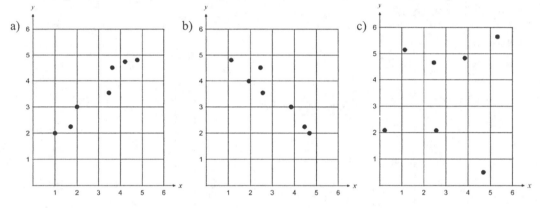

Solution

a) $r \approx 1$; The points have a fairly tight linear relationship with a positive association, with the variables x and y increasing and decreasing together.

b) $r \approx -1$; The points have a strong negative association, with high values of x associated with low values of y, and vice versa.

c) $r \approx 0$; The variables x and y are fluctuating independently, with no clear correlated trend.

☞ Key idea

The following **formula for correlation** can be used given the means and standard deviations of the two variable x and y for the n individuals.

$$r = \frac{1}{n-1}\left[\left(\frac{x_1-\bar{x}}{s_x}\right)\left(\frac{y_1-\bar{y}}{s_y}\right)+\left(\frac{x_2-\bar{x}}{s_x}\right)\left(\frac{y_2-\bar{y}}{s_y}\right)+\dots+\left(\frac{x_n-\bar{x}}{s_x}\right)\left(\frac{y_n-\bar{y}}{s_y}\right)\right]$$

Section 6.4 Least – Squares Regression

☞ Key idea

The **least-squares regression line** runs through a scatterplot of data so as to be the line that makes the sum of the squares of the vertical deviations from the data points to the line as small as possible. This is often thought of as the "line of best fit" to the data.

𝆏 Key idea

The **formula for the equation of the least-square regression line** for a data set on an explanatory variable x and a response variable y depends on knowing the means of x and y, the standard deviations of x and y, and their correlation r. It produces the slope and intercept of the regression line.

The least-square regression line is predicted $y = a + bx$, where $b = r\frac{s_y}{s_x}$ (slope) and $a = \bar{y} - b\bar{x}$ (y-intercept).

⌇ Example D

Given the following data, compute the correlation and least-squares regression line by hand.

x	4	9	3	5	2
y	6	7	4	6	3

Solution

We have the following hand calculations.

i	Observations x_i	Observations y_i	Deviations $x_i - \bar{x}$	Deviations $y_i - \bar{y}$	Squared deviations $(x_i - \bar{x})^2$	Squared deviations $(y_i - \bar{y})^2$
1	4	6	− 0.6	0.8	0.36	0.64
2	9	7	4.4	1.8	19.36	3.24
3	3	4	− 1.6	− 1.2	2.56	1.44
4	5	6	0.4	0.8	0.16	0.64
5	2	3	− 2.6	− 2.2	6.76	4.84
sum	23	26	0	0	29.2	10.8

$\bar{x} = \frac{23}{5} = 4.6$, $\bar{y} = \frac{26}{5} = 5.2$, $s_x^2 = \frac{29.2}{5-1} = \frac{29.2}{4} = 7.3$, $s_x = \sqrt{7.3} \approx 2.7019$, $s_y^2 = \frac{10.8}{5-1} = \frac{10.8}{4} = 2.7$, and $s_y = \sqrt{2.7} \approx 1.6432$.

Since

$$r = \frac{1}{n-1}\left[\left(\frac{x_1 - \bar{x}}{s_x}\right)\left(\frac{y_1 - \bar{y}}{s_y}\right) + \left(\frac{x_2 - \bar{x}}{s_x}\right)\left(\frac{y_2 - \bar{y}}{s_y}\right) + \left(\frac{x_3 - \bar{x}}{s_x}\right)\left(\frac{y_3 - \bar{y}}{s_y}\right) + \left(\frac{x_4 - \bar{x}}{s_x}\right)\left(\frac{y_4 - \bar{y}}{s_y}\right) + \left(\frac{x_5 - \bar{x}}{s_x}\right)\left(\frac{y_5 - \bar{y}}{s_y}\right)\right],$$

we have the following.

$$r \approx \frac{1}{5-1}\left[\left(\frac{-0.6}{2.7019}\right)\left(\frac{0.8}{1.6432}\right) + \left(\frac{4.4}{2.7019}\right)\left(\frac{1.8}{1.6432}\right) + \left(\frac{-1.6}{2.7019}\right)\left(\frac{-1.2}{1.6432}\right)\right.$$

$$\left. + \left(\frac{0.4}{2.7019}\right)\left(\frac{0.8}{1.6432}\right) + \left(\frac{-2.6}{2.7019}\right)\left(\frac{-2.2}{1.6432}\right)\right]$$

$$= \frac{1}{4}\left[\frac{-0.48}{4.4397608} + \frac{7.92}{4.4397608} + \frac{1.92}{4.4397608} + \frac{0.32}{4.4397608} + \frac{5.72}{4.4397608}\right]$$

$$= \frac{1}{4}\left[\frac{15.4}{4.4397608}\right] = \frac{3.85}{4.4397608} \approx 0.8672$$

Since $b = r\frac{s_y}{s_x} = 0.8672\left(\frac{1.6432}{2.7019}\right) \approx 0.527$ and $a = \bar{y} - b\bar{x} = 5.2 - 0.5274(4.6) \approx 2.774$, the least-square regression line is $y = a + bx = 2.774 + 0.527x$.

✎ **Question 2**

Given the following data, compute the correlation and least-squares regression line by hand.

x	1	2	3	4	5
y	8	4	6	5	2

Answer

$r = -0.778$ and $y = 8.3 - 1.1x$.

Section 6.5   Interpreting Correlation and Regression

⌐ **Key idea**

Both the correlation, r, and the least-squares regression line can be strongly influenced by a few outlying points. Never trust a correlation until you have plotted the data.

⌐ **Key idea**

Correlation and regression *describe* relationships. *Interpreting* relationships requires more thought. Try to think about the effects of other variables prior to drawing conclusions when interpreting the results of correlation and regression. An association between variables is not itself good evidence that a change in one variable actually causes a change in the other!

〰 **Example E**

Measure the number of gold rings per women x and the number of deaths from breast cancer y for women of the world's nations. There is a strong correlation: Nations that have woman with many gold rings have fewer deaths from breast cancer. What kind of correlation would this be (negative or positive)? Can woman around the world reduce the number of deaths due to breast cancer by owning rings?

Solution

This should be a negative correlation (called high negative, closer to -1). Women from rich nations should have more gold rings than women from poor nations. Rich nations have better medical treatment for breast cancer and would offer lower death rates as a result. There is no cause-and-effect tie between gold rings and death rates from breast cancer.

✎ **Question 3**

The following is data from a small company. The explanatory variable is the number of years with the company and the response variable is salary. Use a calculator to determine the correlation and least-squares regression line.

x	1 year	2 year	3 year	4 year	5 year
y	$77,500	$29,500	$31,000	$34,000	$41,000

a) Using the regression line, project the salary of an employee that has been with the company 10 years. Comment on the results.

Remove the outlier from the data and compute again the correlation and least-squares regression line.

b) Using the regression line (without the outlier), project the salary of an employee that has been with the company 10 years. Comment on the results.

Answer

a) approximately $-$5350; Comments will vary.

b) approximately $58,250; Comments will vary.

Exercises 40 – 44
Carefully read Section 6.4 before responding to these exercises. Answers will vary in these exercises. Try to think carefully of the potential cause and effect or alternative explanations for the effect.

Exercise 45
Carefully read Section 6.1 before responding to this exercise. Reading section 6.5 may also help in guiding you in interpreting the results.

Exercise 46
Carefully read Section 6.1 before responding to this exercise.

Exercises 47
Carefully read Section 6.3 before responding to this exercise. Make sure you know the course requirements regarding the use of calculators and/or spreadsheets in computing your answers. The following may be helpful in creating the scatterplot for needed this exercise in Part a.

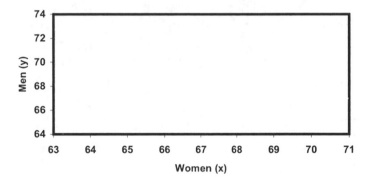

Exercise 48
Carefully read Section 6.3 before responding to this exercise. The section specifically addresses what is asked for in this question.

Exercise 49
Carefully read Section 6.2 before responding to this exercise.

Do You Know the Terms?

Cut out the following 9 flashcards to test yourself on Review Vocabulary. You can also find these flashcards at http://www.whfreeman.com/fapp7e.

Chapter 6 Exploring Data: Relationships **Correlation**	Chapter 6 Exploring Data: Relationships **Intercept of a line**
Chapter 6 Exploring Data: Relationships **Least-squares regression line**	Chapter 6 Exploring Data: Relationships **Negative association**
Chapter 6 Exploring Data: Relationships **Outlier**	Chapter 6 Exploring Data: Relationships **Positive association**
Chapter 6 Exploring Data: Relationships **Regression line**	Chapter 6 Exploring Data: Relationships **Response variable**
Chapter 6 Exploring Data: Relationships **Explanatory variable**	Chapter 6 Exploring Data: Relationships **Scatterplot**

Example 2

Find and graph the least-squares regression line for the following data.

x	2	4	1	5	7
y	6	5	7	7	4

Solution

With data already entered, press the [STAT] button. Toggle to the right for CALC. Toggle down to 8:LinReg(a+bx) and press [ENTER] .

Instead of toggling down to 8:LinReg(a+bx) and pressing [ENTER] , you could alternatively press the 8 button ([8]). In either case the following screen will appear.

By pressing [ENTER] , you may get the following screen. Your screen may have more information.

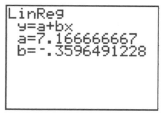

There are several ways to obtain the following graph of the least-squares line along with the scatterplot.

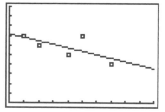

In all three methods, you will need to press [Y=] in order to enter the equation.

Method I: Type in the equation of the regression line, $y = a + bx$, by rounding the values of a and b.

```
Plot1  Plot2  Plot3
\Y1=7.167-.360X
\Y2=
\Y3=
\Y4=
\Y5=
\Y6=
\Y7=
```

Press GRAPH in order to obtain the graph. This is the easiest method.

Method II: Place the equation of the regression line, $y = a + bx$, up to the accuracy of the calculator.

To do this, you press VARS then toggle down to 5:Statistics and press ENTER . You could alternatively press the 5 button (5).

```
VARS  Y-VARS
1:Window...
2:Zoom...
3:GDB...
4:Picture...
5:Statistics...
6:Table...
7:String...
```
```
XY  Σ  EQ  TEST  PTS
1:n
2:x̄
3:Sx
4:σx
5:ȳ
6:Sy
7↓σy
```

Toggle to the right to the EQ menu and press ENTER .

```
XY  Σ  EQ  TEST  PTS
1:RegEQ
2:a
3:b
4:c
5:d
6:e
7↓r
```
```
Plot1  Plot2  Plot3
\Y1=7.1666666666
667+-.3596491228
0702X
\Y2=
\Y3=
\Y4=
\Y5=
```

Press GRAPH in order to obtain the graph.

Method III: Place the equation of the regression line, $y = a + bx$, in general into your equation editor.

To do this, you press VARS .

```
VARS  Y-VARS
1:Window...
2:Zoom...
3:GDB...
4:Picture...
5:Statistics...
6:Table...
7:String...
```

Toggle down to 5:Statistics and press ENTER . You could alternatively press the 5 button (5).

```
XY  Σ  EQ  TEST  PTS
1:n
2:x̄
3:Sx
4:σx
5:ȳ
6:Sy
7↓σy
```

Continued on next page

Toggle to the right to the EQ menu and toggle down to 2:a and press ENTER . You could alternatively press the 2 button (2). After pressing the plus button (+), press VARS again and run through the similar procedure to insert b.

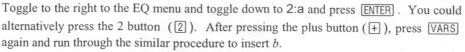

Finally, press X,T,Θ,n to get the following screen.

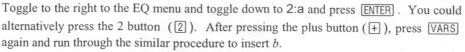

Press GRAPH in order to obtain the graph. Although this is the hardest method, you only need to do it once. When you don't wish for the equation to be graphed, simply de-select the relation by the method described in Chapter 5 of Learning the Calculator.

Example 3

Find the correlation for the following data.

x	2	4	1	5	7
y	6	5	7	7	4

Solution

You may have already obtained the correlation when you obtained the least-squares line in Example 2. If not, you need to activate the DiagnosticOn feature. To do this, press 2nd then 0 . This will take you to the CATALOG menu.

CATALOG
 Degree
 DelVar
 DependAsk
 DependAuto
 det(
 DiagnosticOff
▶DiagnosticOn

Toggle down to DiagnosticOn and press ENTER twice. You should get the following screen.

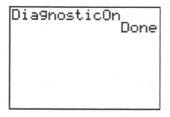

Finally, follow the instructions in Example 2 to obtain the least-squares line. The correlation is $r \approx -0.659$.

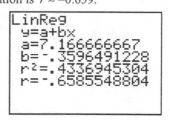

Practice Quiz

1. The park wants to predict the amount of ice used each day, based on the predicted high temperature. Which variable would be the explanatory variable?
   a. the amount of ice used
   b. the predicted high temperature
   c. the actual high temperature

2. The daily ice consumption (in pounds) y at a park is related to the predicted high temperature (in degrees F) x. Suppose the least-squares regression line is $y = 250 + 25x$. Today's predicted high temperature is 90 degrees F. This means that
   a. at least 2500 pounds of ice will be needed today.
   b. approximately 2500 pounds of ice will be needed today.
   c. exactly 2500 pounds of ice will be needed today.

3. The daily ice consumption (in pounds) y at a park is related to the predicted high temperature (in degrees F) x. Suppose the least-squares regression line is $y = 250 + 25x$. Suppose 2300 pounds of ice were used yesterday. This leads you to believe that yesterday's high temperature was closest to
   a. 52 degrees F.
   b. 82 degrees F.
   c. 92 degrees F.

4. Draw a scatterplot showing the relationship between the observed variables x and y, with the data given in the table below.

x	2	4	1	5	7	9	8
y	10	5	8	7	4	6	6

   Based of the scatterplot, you can state that
   a. the y-intercept of the least – squares regression line is about 3.
   b. the correlation, r, is between – 1 and 0.
   c. the slope of the least – squares regression line is about 0.45.

5. Find the equation of the least – squares regression line for the data below.

x	2	4	5	2	3	6	7	1
y	5	6	6	1	5	8	7	3

   a. $y = 1.88 + 0.865x$
   b. $y = 1.88 - 0.865x$
   c. $y = 2.54 + 0.865x$

6. Suppose you collected data for the number of hours individuals practice bowling, x, versus their bowling score, y. What is the best conclusion that can be drawn?
   a. The correlation should be close to zero.
   b. The correlation should be positive.
   c. The correlation should be negative.

Chapter 7
Data for Decisions

Chapter Objectives

Check off these skills when you feel that you have mastered them.

☐ Identify the population in a given sampling or experimental situation.

☐ Identify the sample in a given sampling or experimental situation.

☐ Explain the difference between a population and a sample.

☐ Analyze a sampling example to detect sources of bias.

☐ Identify several examples of sampling that occur in our society.

☐ Select a numbering scheme for a population from which a random sample will be selected and use a table of random digits to select that random sample.

☐ Explain the difference between an observational study and an experiment.

☐ Recognize the confounding on the effects of two variables in an experiment.

☐ Explain the difference between the experimental group and the control group in an experiment.

☐ Design a randomized comparative experiment and display it in graphical form.

☐ Explain what is meant by statistically significant.

☐ Describe the placebo effect.

☐ Discuss why double blindness is desirable in an experiment.

☐ Define statistical inference.

☐ Explain the difference between a parameter and a statistic.

☐ Identify both the parameter and the statistic in a simple inferential setting.

☐ Compute the sample proportion when both the sample size and number of favorable responses are given.

☐ Using an appropriate formula, calculate the standard deviation of a given statistic.

☐ Explain the difference between the population mean and the sample mean.

☐ Given a sample proportion and sample size, list the range for a 95% confidence interval for the population proportion.

☐ Calculate differing margins of error for increasing sample sizes.

☐ Discuss the effect of an increased sample size on the statistic's margin of error.

Guided Reading

Introduction

Numbers are used in a myriad of ways to describe the world we live in. Our social and economic concerns, science and ecology, politics, religion, health, recreation, any area of human activity is better understood by the collection and analysis of data. It is vitally important to know how to produce trustworthy data, and how to draw reliable conclusions from them. This is the role of statistics, the science of data handling.

☞ Key idea

In Chapters 5 and 6, **data analysis** was explored. Graphs and numbers were produced to represent a set of data. In this chapter you will explore **how to produce data** that can be trusted for answering specific questions. Then in turn answering questions with a certain degree of confidence is **statistical inference**.

Section 7.1   Sampling

☞ Key idea

In statistical studies, we gather information about a small, partial group (a **sample**) in order to draw conclusions about the whole, large group we are interested in (the **population**).

⌇ Example A

In a study of the smoking habits of urban American adults, we ask 500 people their age, place of residence, and how many cigarettes they smoke daily. What is the population and what is the sample?

Solution

The population is all Americans who live in a large city and are old enough to be classified as adults. The sample consists of those among the 500 interviewed who qualify as members of this population; for example, children would not be included, even if they smoke.

Section 7.2   Bad Sampling Methods

☞ Key idea

Systematic error caused by bad sampling methods may lead to a **biased** study favoring certain outcomes.

⌇ Example B

Customers at a supermarket are sampled to determine their opinion about a volatile political issue. Can you identify a possible source of bias in such a survey?

Solution

There may be gender-based bias, with women overrepresented in the sample. Depending on the location of the market, there may also be a bias according to economic class, education level, political affiliation, etc.

☞ Key idea

A sample of people who choose to respond to a general appeal is called a **voluntary response sample**. This voluntary response is a likely source of bias.

ᕦ Example C

Television viewers are invited to call an 800 number to report their opposition to a bill to increase state gasoline taxes. Why might this survey be biased?

Solution

There is a high likelihood that a disproportionately large number of people angry about a potential tax increase will take the trouble to call to register their opposition.

✐ Question 1

Consider the following.

       a.   convenience sample   b.   voluntary response sample   c.   bias

Which of the above expressions/word would **not** be used to fill in a blank for the following?

1)   To determine the food preferences of students, a staff member surveys students as they exit a local bar. This type of sample is a _____.

2) A survey on the benefits of jogging is conducted outside a sporting-goods store. This is an example of _____.

Answer

b

Section 7.3   Simple Random Samples

⚷ Key idea

We can use a **simple random sample (SRS)** to eliminate bias. This is the equivalent of choosing names from a hat; each individual has an equal chance to be selected.

ᕦ Example D

To choose a sample of five cards from a deck of 52 cards, you shuffle the cards and choose the first, third, fifth, seventh, and ninth card. Will this lead to a simple random sample?

Solution

Yes, if the deck has been thoroughly shuffled. After shuffling, any given card is equally likely to occupy any given position in the deck.

⚷ Key idea

A two-step procedure for forming a SRS using a table of random digits is:

Step 1: Give each member a numerical label of the same length.

Step 2: Read from the table strings of digits of the same length as the labels. Ignore groups not used as labels and also ignore any repeated labels.

ᕦ Example E

Describe how to use the table of random digits to form a random sample of 75 students at Hypothetical University from the entire population of 1350 HU students.

Solution

Assign each HU student a four digit numerical label, 0001–1350, making sure that no label is assigned twice. Then starting anywhere in the random digit table, read strings of 4 consecutive digits, ignoring repetitions and unassigned strings, until 75 assigned labels are obtained. The students with those labels are the sample.

⌐ Key idea

A **statistic** is a number that described a **sample**. This value can change from sample to sample. A statistic is often used to estimate an unknown parameter.

✐ Question 3

Consider the following.

       a. statistic   b. sample   c. parameter

Which of the above expressions/word would **not** be used to fill in a blank for the following?

A random sample of 10 bags of flour has a mean weight of 24.9 pounds, less than the mean weight 25.05 pounds of all bags produced.

1)   In this example. 25.05 is called a _____.
2)   In this example, 24.9 is called a _____.

Answer

b

⌐ Key idea

If you have a simple random sample of size n from a large population and a count of success (such as agreeing with a survey question) in the same population then the **sample proportion of successes**, $\left(\hat{p}\right)$, is the following quotient.

$$\hat{p} = \frac{\text{count of successes in sample}}{n}$$

$\hat{p}$ is a statistic. The corresponding population proportion parameter is p.

⌐ Key idea

Results of a survey will vary from sample to sample. The margin of error given for a national sample indicates how close that result is to the truth. If the population result would fall in 95% of all samples drawn using the same method, we say we have 95% confidence that the truth about the population falls within this margin of error.

ᑈ Example K

A random sample of 150 people are asked if they own dogs, and 58 of them say yes. What would you estimate the percentage of dog owners to be in the general population?

Solution

The sample proportion is $\hat{p} = \frac{58}{150} \approx 0.387 = 38.7\%$, the actual population proportion may differ somewhat, but is reasonably likely to be fairly close to that of the sample. Thus, our best estimate is 38.7%.

✐ Question 4

Suppose you conduct a telephone poll of 1250 people, asking them whether or not they favor mandatory sentencing for drug related crimes. If 580 people say "yes," what is the sample proportion $\hat{p}$ of people in favor of mandatory sentencing?

Answer

$\hat{p} = 46.4\%$

ᘒ Example C

Television viewers are invited to call an 800 number to report their opposition to a bill to increase state gasoline taxes. Why might this survey be biased?

Solution

There is a high likelihood that a disproportionately large number of people angry about a potential tax increase will take the trouble to call to register their opposition.

✐ Question 1

Consider the following.

      a.   convenience sample   b.   voluntary response sample   c.   bias

Which of the above expressions/word would **not** be used to fill in a blank for the following?

1) To determine the food preferences of students, a staff member surveys students as they exit a local bar. This type of sample is a _____.

2) A survey on the benefits of jogging is conducted outside a sporting-goods store. This is an example of _____.

Answer

b

Section 7.3   Simple Random Samples

ᘁ Key idea

We can use a **simple random sample** (**SRS**) to eliminate bias. This is the equivalent of choosing names from a hat; each individual has an equal chance to be selected.

ᘒ Example D

To choose a sample of five cards from a deck of 52 cards, you shuffle the cards and choose the first, third, fifth, seventh, and ninth card. Will this lead to a simple random sample?

Solution

Yes, if the deck has been thoroughly shuffled. After shuffling, any given card is equally likely to occupy any given position in the deck.

ᘁ Key idea

A two-step procedure for forming a SRS using a table of random digits is:

Step 1:   Give each member a numerical label of the same length.

Step 2:   Read from the table strings of digits of the same length as the labels. Ignore groups not used as labels and also ignore any repeated labels.

ᘒ Example E

Describe how to use the table of random digits to form a random sample of 75 students at Hypothetical University from the entire population of 1350 HU students.

Solution

Assign each HU student a four digit numerical label, 0001–1350, making sure that no label is assigned twice. Then starting anywhere in the random digit table, read strings of 4 consecutive digits, ignoring repetitions and unassigned strings, until 75 assigned labels are obtained. The students with those labels are the sample.

⬿ Example F

A teacher wants to randomly poll her students regarding whether they liked a certain project or not. There are 25 students in the class and he wants to poll five of them. Starting at line 105 use the partial table (Table 7.1 of your text) to find the five random students.

Adam	Faiz	Kevin	Patty	Victoria
Billy	Gwen	Leo	Quinn	Wally
Cassy	Heidi	Mary	Rachel	Xavier
Daniel	Iliana	Nadia	Sarah	Yaffa
Edwin	Jacob	Ottis	Thomas	Zeki

TABLE 7.1   Random Digits

101	19223	95034	05756	28713	96409	12531	42544	82853
102	73676	47150	99400	01927	27754	42648	82425	36290
103	45467	71709	77558	00095	32863	29485	82226	90056
104	52711	38889	93074	60227	40011	85848	48767	52573
105	95592	94007	69971	91481	60779	53791	17297	59335
106	68417	35013	15529	72765	85089	57067	50211	47487
107	82739	57890	20807	47511	81676	55300	94383	14893
108	60940	72024	17868	24943	61790	90656	87964	18883
109	36009	19365	15412	39638	85453	46816	83485	41979
110	38448	48789	18338	24697	39364	42006	76688	08708
111	81486	69487	60513	09297	00412	71238	27649	39950
112	59636	88804	04634	71197	19352	73089	84898	45785
113	62568	70206	40325	03699	71080	22553	11486	11776
114	45149	32992	75730	66280	03819	56202	02938	70915

Solution

Step 1: Give each student a label of the same numerical length.

01 Adam	06 Faiz	11 Kevin	16 Patty	21 Victoria
02 Billy	07 Gwen	12 Leo	17 Quinn	22 Wally
03 Cassy	08 Heidi	13 Mary	18 Rachel	23 Xavier
04 Daniel	09 Iliana	14 Nadia	19 Sarah	24 Yaffa
05 Edwin	10 Jacob	15 Ottis	20 Thomas	25 Zeki

Step 2: Use the table starting on line 105 looking at groups of digits of length 2.

TABLE 7.1   Random Digits

101	19223	95034	05756	28713	96409	12531	42544	82853
102	73676	47150	99400	01927	27754	42648	82425	36290
103	45467	71709	77558	00095	32863	29485	82226	90056
104	52711	38889	93074	60227	40011	85848	48767	52573
105	95592	94007	69971	91481	60779	53791	17297	59335
106	68417	35013	15529	72765	85089	57067	50211	47487
107	82739	57890	20807	47511	81676	55300	94383	14893
108	60940	72024	17868	24943	61790	90656	87964	18883
109	36009	19365	15412	39638	85453	46816	83485	41979
110	38448	48789	18338	24697	39364	42006	76688	08708
111	81486	69487	60513	09297	00412	71238	27649	39950
112	59636	88804	04634	71197	19352	73089	84898	45785
113	62568	70206	40325	03699	71080	22553	11486	11776
114	45149	32992	75730	66280	03819	56202	02938	70915

Thus, the students she would poll are 07 = Gwen, 19 = Sarah, 14 = Nadia, 17 = Quinn, and 13 = Mary.

✏ Question 2

Redo Example F, but start on line 102. In alphabetical order, who would be the third student that the teacher would select to poll?

Answer

Quinn

Section 7.4   Cautions about Sample Surveys

⌐ Key idea

Even a sound statistical design cannot guard against some of the pitfalls associated with statistical experiments. For example, **nonresponse** can be a cause of bias in an experiment, as can the artificial environments created for some experiments and **undercoverage**, by not including in samples certain parts of the population. Responses can be strongly influenced by the **wording of questions**. By having leading questions or confusing questions, strong bias can be introduced.

↝ Example G

The Highway Patrol in a state decides to estimate the average speed of drivers using the fast lane. Using their patrol cars, they get behind them and record their speed which would be the same as the car in front of them. Is there any bias?

Solution

Yes, there is bias. Most likely the drives will slow down upon seeing the patrol car. This will result in overall lower average speed.

Section 7.5   Experiments

⌐ Key idea

An **observational study** is a passive study of a variable of interest. The study does not attempt to influence the responses and is meant to *describe* a group or situation.

⌐ Key idea

An **experiment** is an active trial of an imposed *treatment* and its *effects*. The study is meant to observe whether the treatment causes a change in the response.

↝ Example H

Which is an experiment and which is an observational study?
a)   You ask a sample of smokers how many cigarettes they smoke daily, and measure their blood pressure.
b)   You select a sample of smokers and measure their blood pressure. Then you ask them to reduce their smoking by 5 cigarettes a day; after 3 months you recheck their blood pressure.

Solution

a)   This is an observational study. We are passively observing and measuring.
b)   This is an experiment. We are actively influencing the behavior of the subjects.

☞ **Key idea**

When designing an **uncontrolled study**, care must be taken to avoid **confounded variables**. Confounding variables are variables whose effects on the outcome cannot be distinguished from one another.

☞ **Key idea**

We can reduce the effect of confounded variables by conducting a **randomized comparative experiment**. The sample for the experiment is matched by a **control group**, with subjects assigned randomly to the treatment or the control group. Since personal choice can be a source of bias, the subjects should be randomly chosen for each group.

& **Example I**

How would you design a simple randomized comparative experiment to test the effect of a high-potassium diet on smokers' blood pressure? Assume you have 200 smokers who have agreed to participate in the experiment.

Solution

From the group of 200 smokers who have agreed to participate, randomly select 100 to try the high-potassium diet. The other 100 will serve as the control group, and will make no change in their diet. Measure the blood pressure of each subject at the beginning and end of the testing period, and compare changes in the two groups.

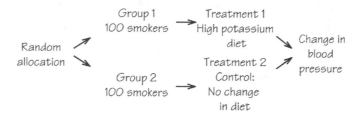

☞ **Key idea**

A well-designed experiment is one that uses the principles of **comparison** and **randomization**: comparison of several treatments and the random assignment of subjects to treatments.

☞ **Key idea**

If subjects are randomly assigned to treatments, we can be confident that any differences among treatment groups that are too large to have occurred by chance are **statistically significant**. Small differences between groups in a study can be due to random variation, but statistically significant differences are too large to be attributable to chance and are reliable evidence of a real effect of the factors being studied.

Section 7.6   Thinking about Experiments

⌇→ **Key idea**

The **placebo effect** is a special kind of confounding in which a patient responds favorably to any treatment, even a placebo (fake treatment).

⌇→ **Key idea**

To avoid the placebo effect and any possible bias on the part of the experimenters, use a **double-blind** experiment, so neither subjects nor investigators know which treatment an individual is receiving.

∾ **Example J**

How would you design a double-blind experiment to test the effect of a vitamin supplement on smokers' blood pressure?

Solution

Randomly assign labels 001 – 200 to your subjects. Using the labels, randomly choose 100 of the subjects to receive the vitamin supplement (say, in pill form), while the other group receives an indistinguishable placebo. The list of which group each subject belongs to is kept confidential; neither subjects nor experimenters know who is taking the real supplement until the experiment is over and the data have been recorded. This way, neither psychological factors nor unconscious bias on the part of the experimenters can play a role.

⌇→ **Key idea**

A **prospective study** is an observational study that records slowly developing effects of a group of subjects over a long period of time.

⌇→ **Key idea**

Only experimentation can produce fully convincing statistical evidence of **cause and effect**.

⌇→ **Key idea**

Experiments like samples have weaknesses, in particular, they can **lack realism**. This would mean that it is hard to say exactly how far the results of the experiment can be applied.

Section 7.7   Inference: From Sample to Population

⌇→ **Key idea**

Using a fact about a sample to estimate the truth about the whole population is called **statistical inference**. We are inferring conclusions about the whole population based on data from selected individuals. Statistical inference only works if the data comes from a random sample or a randomized comparative experiment. A sample should resemble the population, so that a sample statistic can be used to estimate a characteristic of the population.

⌇→ **Key idea**

A **parameter** is a number that describes the **population**. A parameter is a fixed value, but we generally do not know what it is.

⌐ Key idea

A **statistic** is a number that described a **sample**. This value can change from sample to sample. A statistic is often used to estimate an unknown parameter.

✎ Question 3

Consider the following.

                    a. statistic   b. sample   c. parameter

Which of the above expressions/word would **not** be used to fill in a blank for the following?

A random sample of 10 bags of flour has a mean weight of 24.9 pounds, less than the mean weight 25.05 pounds of all bags produced.

1)   In this example. 25.05 is called a _____.
2)   In this example, 24.9 is called a _____.

Answer
b

⌐ Key idea

If you have a simple random sample of size n from a large population and a count of success (such as agreeing with a survey question) in the same population then the **sample proportion of successes**, $\left(\hat{p}\right)$, is the following quotient.

$$\hat{p} = \frac{\text{count of successes in sample}}{n}$$

$\hat{p}$ is a statistic. The corresponding population proportion parameter is p.

⌐ Key idea

Results of a survey will vary from sample to sample. The margin of error given for a national sample indicates how close that result is to the truth. If the population result would fall in 95% of all samples drawn using the same method, we say we have 95% confidence that the truth about the population falls within this margin of error.

ᕦ Example K

A random sample of 150 people are asked if they own dogs, and 58 of them say yes. What would you estimate the percentage of dog owners to be in the general population?

Solution

The sample proportion is $\hat{p} = \frac{58}{150} \approx 0.387 = 38.7\%$, the actual population proportion may differ somewhat, but is reasonably likely to be fairly close to that of the sample. Thus, our best estimate is 38.7%.

✎ Question 4

Suppose you conduct a telephone poll of 1250 people, asking them whether or not they favor mandatory sentencing for drug related crimes. If 580 people say "yes," what is the sample proportion $\hat{p}$ of people in favor of mandatory sentencing?

Answer
$\hat{p} = 46.4\%$

⟝ Key idea

Statistical inference is based on the idea that one needs to see how trustworthy a procedure is if it is repeated many times. Results of a survey will vary from sample to sample; this is called sampling variability. So to answer the question as to what would happen for many samples, we do the following.

- Take a large number of random samples from the same population.
- Calculate $\hat{p}$ for each sample.
- Make a histogram of $\hat{p}$.
- Examine the distribution for shape, center, and spread, as well as outliers or other deviations.

ᝮ Example L

Suppose the previous dog survey was conducted simultaneously by twelve investigators, each sampling 150 people, leading to the following twelve percentages of dog owners: {39%, 37%, 37%, 39%, 40%, 38%, 41%, 40%, 39%, 41%, 42%, 39%}. Sketch a histogram for this data and discuss the features of the distribution.

Solution

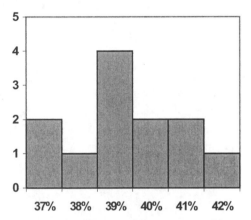

A sampling histogram will generally display a regular pattern with two important features: The results will be centered symmetrically around a peak, the true population value. The spread of the data will be tighter for large sample sizes, wider for small ones.

⟝ Key idea

The **sampling distribution** of a statistic is the distribution of values taken by the statistic in all possible samples of the same size from the same population. This is the ideal pattern if we looked at all same-size possible samples of the population. For a simple random sample of size n from a large population that contains population proportion p the sampling distribution for $\hat{p}$ is approximately normal, with mean p, and standard deviation $\sqrt{\dfrac{p(1-p)}{n}}$. The formula for the standard deviation of $\hat{p}$ shows that the spread of the sampling distribution is about the same for most sample proportions; it depends primarily on the sample size.

ᨆ Example M

Suppose 24% of all college students think that textbook prices are reasonable. If you take a random sample of 2000 college students, what is the standard deviation of $\hat{p}$? Round to four decimal places.

Solution

Convert 24% to decimal form, 0.24. Standard deviation is as follows.

$$\sqrt{\frac{p(1-p)}{n}} = \sqrt{\frac{0.24(1-0.24)}{2000}} = \sqrt{\frac{0.24(0.76)}{2000}} \approx 0.0095$$

✎ Question 5

Suppose that in the political poll from Question 4, the true population proportion is $p = 45\%$. What is the standard deviation of the sampling distribution?

Answer

The standard deviation of $\hat{p}$ is approximately 0.0141.

Section 7.8   Confidence Intervals

ᨆ Key idea

We cannot know precisely a true population parameter, such as the proportion p of people who favor a particular political candidate. To make an estimate, we interview a random sample of the population and calculate a statistic of the sample, such as the sample proportion, $\hat{p}$, favoring the candidate in question. Since $\hat{p}$ is close to normal in its distribution, we will consider the 95 part of the 68-95-99.7 rule which indicates that 95% of all samples of $\hat{p}$ will fall within two standard deviations of the true population proportion, p. This leads us to the **95% confidence interval for p**, which is quite accurate for large values of n. It is as follows.

$$\hat{p} \pm 2\sqrt{\frac{\hat{p}(1-\hat{p})}{n}}$$

ᨆ Key idea

The **margin of error** of a survey gives an interval that includes 95% of the samples and is centered around the true population value. The margin of error is $2\sqrt{\frac{\hat{p}(1-\hat{p})}{n}}$.

ᨆ Example N

Suppose the results of the dog survey are announced as follows: "The percentage of people who own dogs is 39%, with a margin of error of 4%". Can you say the following with reasonable (about 95%) certainty of being right?
a)   At least (that is, not less than) 39% of people own dogs.
b)   At most (that is, not more than) 45% of people own dogs.

Solution

a)   No. The true percentage of dog owners is just as likely to be below 39% as above it.
b)   Yes. With a 4% margin of error, the true percentage is almost certain to be within the 35%–43% range, and is thus highly likely to be less than 45%.

↶ Example O

In a political poll, 695 potential voters are asked if they have decided yet which candidate they will vote for in the next election. Suppose that 511 say "yes."

a) Estimate the proportion of *undecided* voters.

b) Find a 95% confidence interval for this estimate.

Solution

a) The number of voters in the sample who have made up their minds is 511. Thus, the number of undecided voters in the sample is $695 - 511 = 184$. The sample proportion of voters is therefore as follows.

$$\hat{p} = \frac{184}{695} \approx 0.265 = 26.5\%$$

b) The 95% confidence interval for this estimate can be calculated as follows.

$$\hat{p} \pm 2\sqrt{\frac{\hat{p}(1-\hat{p})}{n}} = 0.265 \pm 2\sqrt{\frac{0.265(1-0.265)}{695}} = 0.265 \pm 2\sqrt{\frac{0.265(0.735)}{695}} \approx 0.265 \pm 0.033$$

$$0.265 - 0.033 = 0.232 = 23.2\% \text{ to } 0.265 + 0.033 = 0.298 = 29.8\%$$

Rounding off we get an interval of $(23\%, 30\%)$.

✎ Question 6

In a college survey, 847 students were asked if they thought the cost of tuition ias reasonable. 521 said that they felt it was reasonable.

a) Estimate the proportion of students that believe the cost of tuition is reasonable.

b) Find a 95% confidence interval for this estimate.

Answer

a) 61.5%

b) (58.2%, 64.8%)

Homework Help

☞ To assist you in your homework, a copy of Table 7.1 appears after this section.

Exercises 1 – 2
Carefully read Section 7.1 before responding to these exercises.

Exercises 3 – 6
Carefully read Section 7.2 before responding to these exercises. Your answers may differ from a classmate in terms of describing reasons for bias or giving examples. Try to imagine yourself in the situation described in the exercise before responding.

Exercises 7 – 10 & 13
Carefully read Section 7.3 before responding to these exercises. You may choose to make some copies of Table 7.1 to write on. Be very careful using the table by taking your time. Take the time to check your answer twice. If you write on the table in pencil and plan to reuse it, make sure to erase any stray marks so that they will not interfere with using the table again.

Exercises 11 – 12
Look carefully at Table 7.1 as you respond to these exercises.

Exercise 14
Carefully read Section 7.3 before responding to this exercise. Pay particular attention to the definition of a simple random sample.

Exercises 15 – 17
Carefully read Section 7.4 before responding to these exercises. Read the scenario in each question carefully and try to imagine yourself in the situation described in an exercise before responding.

Exercises 18 – 27 & 29 – 30
Carefully read Section 7.5 before responding to these exercises. You will need Table 7.1 for Exercises 23, 24, 26, 27, 29, and 30. For Exercises 22, 23, 24, and 29, you will need to draw using the following as a template.

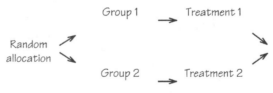

For Exercise 25, you will need to draw using the following as a template.

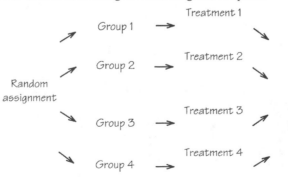

For Exercise 27, you will need to draw using the following as a template.

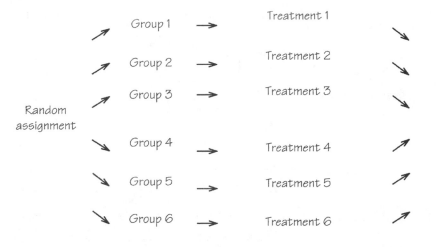

Exercises 28 & 31 – 35
Carefully read Section 7.6 before responding to these exercises. You will need Table 7.1 for **Exercise** 31. For Exercise 31, you will need to draw using the following as a template.

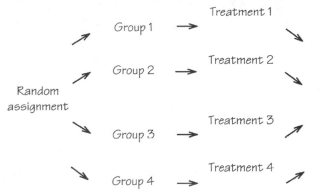

Exercises 36 – 41
Carefully read Section 7.7 before responding to these exercises. Look carefully at Table 7.1 **as you** respond to Exercise 41. You will need a calculator with the square root feature for Exercises 38 – **40**. Make sure you know the requirements as to how much work should be shown (including steps **for** rounding) for your homework.

Exercises 42 – 53
Carefully read Section 7.8 before responding to these exercises. You will need a calculator with **the** square root feature for Exercises 42 – 46. Make sure you know the requirements as to how **much** work should be shown (including steps for rounding) for your homework. In Exercise 52, if we let E

be the margin of error, then we have $E = 2\sqrt{\dfrac{\hat{p}\left(1-\hat{p}\right)}{n}}$. Consider what happens to this formula **when**

you need to find $\dfrac{E}{2}$.

Exercise 54

Carefully read Section 7.5 before responding to this exercise. You will need to draw using the following as a template.

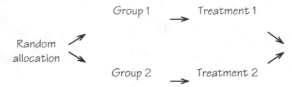

Exercise 55

Carefully read Section 7.8 before responding to this exercise.

Exercise 56

Carefully read Sections 7.3 and 7.4 before responding to this exercise. You will need Table 7.1 for this exercise.

Exercise 57

Carefully read Section 7.7 before responding to this exercise. Also, recall the 68-95-99.7 rule as you work this exercise.

TABLE 7.1   Random   Digits

101	19223	95034	05756	28713	96409	12531	42544	82853
102	73676	47150	99400	01927	27754	42648	82425	36290
103	45467	71709	77558	00095	32863	29485	82226	90056
104	52711	38889	93074	60227	40011	85848	48767	52573
105	95592	94007	69971	91481	60779	53791	17297	59335
106	68417	35013	15529	72765	85089	57067	50211	47487
107	82739	57890	20807	47511	81676	55300	94383	14893
108	60940	72024	17868	24943	61790	90656	87964	18883
109	36009	19365	15412	39638	85453	46816	83485	41979
110	38448	48789	18338	24697	39364	42006	76688	08708
111	81486	69487	60513	09297	00412	71238	27649	39950
112	59636	88804	04634	71197	19352	73089	84898	45785
113	62568	70206	40325	03699	71080	22553	11486	11776
114	45149	32992	75730	66280	03819	56202	02938	70915
115	61041	77684	94322	24709	73698	14526	31893	32592
116	14459	26056	31424	80371	65103	62253	50490	61181
117	38167	98532	62183	70632	23417	26185	41448	75532
118	73190	32533	04470	29669	84407	90785	65956	86382
119	95857	07118	87664	92099	58806	66979	98624	84826
120	35476	55972	39421	65850	04266	35435	43742	11937
121	71487	09984	29077	14863	61683	47052	62224	51025
122	13873	81598	95052	90908	73592	75186	87136	95761
123	54580	81507	27102	56027	55892	33063	41842	81868
124	71035	09001	43367	49497	72719	96758	27611	91596
125	96746	12149	37823	71868	18442	35119	62103	39244
126	96927	19931	36809	74192	77567	88741	48409	41903
127	43909	99477	25330	64359	40085	16925	85117	36071
128	15689	14227	06565	14374	13352	49367	81982	87209
129	36759	58984	68288	22913	18638	54303	00795	08727
130	69051	64817	87174	09517	84534	06489	87201	97245
131	05007	16632	81194	14873	04197	85576	45195	96565
132	68732	55259	84292	08796	43165	93739	31685	97150
133	45740	41807	65561	33302	07051	93623	18132	09547
134	27816	78416	18329	21337	35213	37741	04312	68508
135	66925	55658	39100	78458	11206	19876	87151	31260
136	08421	44753	77377	28744	75592	08563	79140	92454
137	53645	66812	61421	47836	12609	15373	98481	14592
138	66831	68908	40772	21558	47781	33586	79177	06928
139	55588	99404	70708	41098	43563	56934	48394	51719
140	12975	13258	13048	45144	72321	81940	00360	02428
141	96767	35964	23822	96012	94591	65194	50842	53372
142	72829	50232	97892	63408	77919	44575	24870	04178
143	88565	42628	17797	49376	61762	16953	88604	12724
144	62964	88145	83083	69453	46109	59505	69680	00900
145	19687	12633	57857	95806	09931	02150	43163	58636
146	37609	59057	66967	83401	60705	02384	90597	93600
147	54973	86278	88737	74351	47500	84552	19909	67181
148	00694	05977	19664	65441	20903	62371	22725	53340
149	71546	05233	53946	68743	72460	27601	45403	88692
150	07511	88915	41267	16853	84569	79367	32337	03316

Do You Know the Terms?

Cut out the following 25 flashcards to test yourself on Review Vocabulary. You can also find these flashcards at http://www.whfreeman.com/fapp7e.

Chapter 7 **Data for Decisions** Bias	**Chapter 7** **Data for Decisions** 95% confidence interval
Chapter 7 **Data for Decisions** Confounding	**Chapter 7** **Data for Decisions** Control group
Chapter 7 **Data for Decisions** Convenience sample	**Chapter 7** **Data for Decisions** Double-blind experiment
Chapter 7 **Data for Decisions** Experiment	**Chapter 7** **Data for Decisions** Margin of error
Chapter 7 **Data for Decisions** Nonresponse	

An interval computed from a sample by a method that captures the unknown parameter in 95% of all possible samples. When we calculate the interval for a single sample, we are 95% confident that the interval captures the unknown parameter.

A systematic error that tends to cause the observations to deviate in the same direction from the truth about the population whenever a sample or experiment is repeated.

A group of experimental subjects who are given a standard treatment or no treatment (such as a placebo).

Two variables are confounded when their effects on the outcome of a study cannot be distinguished from each other.

An experiment in which neither the experimental subjects nor the persons who interact with them know which treatment each subject received.

A sample that consists of the individuals who are most easily available, such as people passing by in the street. A convenience sample is usually biased.

As announced by most national polls, the margin of error says how close to the truth about the population the sample result would fall in 95% of all samples drawn by the method used to draw this one sample.

A study in which treatments are applied to people, animals, or things in order to observe the effect of the treatments.

Some individuals chosen for a sample cannot be contacted or refuse to participate.

Chapter 7 Data for Decisions **Observational study**	Chapter 7 Data for Decisions **Parameter**
Chapter 7 Data for Decisions **Placebo effect**	Chapter 7 Data for Decisions **Population**
Chapter 7 Data for Decisions **Prospective study**	Chapter 7 Data for Decisions **Randomized comparative experiment**
Chapter 7 Data for Decisions **Sample**	Chapter 7 Data for Decisions **Sample proportion**

A number that describes the population. In statistical inference, the goal is often to estimate an unknown parameter or make a decision about its value.	A study (such as a sample survey) that observes individuals and measures variables of interest but does not attempt to influence the responses.
The entire group of people or things that we want information about.	The effect of a dummy treatment (such as an inert pill in a medical experiment) on the response of subjects.
An experiment to compare two or more treatments in which people, animals, or things are assigned to treatments by chance.	An observational study that follows two or more groups of subjects forward in time.
The proportion $\hat{p}$ of the members of a sample having some characteristic (such as agreeing with an opinion poll question). The sample proportion from a simple random sample is used to estimate the corresponding proportion pin the population from which the sample was drawn.	A part of the population that is actually observed and used to draw conclusions, or inferences, about the entire population.

Chapter 7 Data for Decisions **Sampling distribution**	Chapter 7 Data for Decisions **Simple random sample (SRS)**
Chapter 7 Data for Decisions **Statistic**	Chapter 7 Data for Decisions **Statistical inference**
Chapter 7 Data for Decisions **Statistical significance**	Chapter 7 Data for Decisions **Table of random digits**
Chapter 7 Data for Decisions **Undercoverage**	Chapter 7 Data for Decisions **Voluntary response sample**

A sample chosen by chance, so that every possible sample of the same size has an equal chance to be the one selected.

The distribution of values taken by a statistic when all possible random samples of the same size are drawn from the same population. The sampling distributions of sample proportions are approximately normal.

Methods for drawing conclusions about an entire population on the basis of data from a sample. Confidence intervals are one type of this method.

A number that describes a sample. A statistic can be calculated from the sample data alone; it does not involve any unknown parameters of the population.

A table whose entries are the digits 0, 1, 2, 3, 4, 5, 6, 7, 8, 9 in a completely random order. That is, each entry is equally likely to be any of the 10 digits and no entry gives information about any other entry.

An observed effect is statistically significant if it is so large that it is unlikely to occur just by chance in the absence of a real effect in the population from which the data were drawn.

A sample of people who choose themselves by responding to a general invitation to give their opinions. Such a sample is usually strongly biased.

The process of choosing a sample may systematically leave out some groups in the population, such as households without a television.

Learning the Calculator

Example 1

Calculate the standard deviation of $\hat{p}$ given that $p = 0.125$ and $n = 2153$.

Solution

Since $\sqrt{\dfrac{p(1-p)}{n}} = \sqrt{\dfrac{0.125(1-0.125)}{2153}}$, we can enter the following into the calculator.

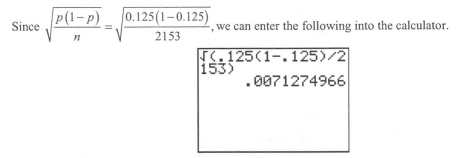

Thus, the standard deviation is approximately 0.007.

Example 2

In a political poll, 10134 potential voters are asked if they have decided yet which candidate they will vote for in the next election. Suppose that 6215 say "yes."
a)   Estimate the proportion of decided voters.
b)   Find a 95% confidence interval for this estimate.

Solution

a)   The sample proportion of voters is therefore as follows.

$$\hat{p} = \frac{6215}{10{,}134}$$

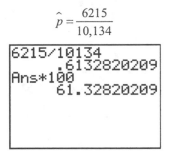

$\hat{p}$ would be approximately $0.613 = 61.3\%$

Continued on next page

b)   The 95% confidence interval for this estimate can be calculated as follows.

$$\hat{p} \pm 2\sqrt{\frac{\hat{p}(1-\hat{p})}{n}} = 0.613 \pm 2\sqrt{\frac{0.613(1-0.613)}{10,134}}$$

Calculate $2\sqrt{\dfrac{0.613(1-0.613)}{10,134}}$ first.

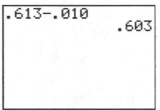

```
2√(.613(1-.613)/
10134)
          .0096766552
```

The 95% confidence interval for this estimate is 0.613 ± 0.010.   First calculate $0.613 - 0.010$.

```
.613-.010
               .603
```

To calculate $0.613 + 0.010$, you can save some keystrokes by pressing [2nd] then [ENTER] and then edit by using the [◄].   Change the – to a +. Then press [ENTER] .

```
.613-.010
               .603
.613+.010
               .623
```

The 95% confidence interval for this estimate is $(60.3\%, 62.3\%)$.

Practice Quiz

1. A marketing firm interviewed 80 shoppers randomly selected from the 4000 customers at one of the mall's 45 stores yesterday. The sample in this situation is the

    a. 45 stores.

    b. 80 selected shoppers.

    c. 4000 customers.

2. In an election for mayor, there are 3 candidates and 24,000 eligible voters. A newspaper interviews 240 voters as they come out of the polls. The population here is the

    a. 240 voters interviewed.

    b. 3 candidates.

    c. 24,000 eligible voters.

3. A well-designed survey should minimize

    a. bias.

    b. randomness.

    c. the placebo effect.

4. Here is a list of random numbers: 16807 64853 17463 14715. Use this list to choose 3 numbers from the set $\{1, 2, 3, \ldots, 20\}$. What are the numbers?

    a. 16, 7, 17

    b. 16, 17, 15

    c. 16, 17, 14

5. If Adam's sample statistic has a margin of error of 3% and Nadia's sample statistic has a margin of error of 5%, then

    a. Nadia's estimate is biased.

    b. Adam has more samples.

    c. Nadia's experiment gave a higher sample estimate.

6. Five workers each sample 50 students to determine their favorite fast-food restaurant. Each worker returns with slightly different results. This is probably due to

    a. bias.

    b. sampling variability.

    c. the use of a control group.

7. A drug test randomly selects one of three treatments for each participant.   Neither the experimenter nor the participant knows which drug is chosen. This is an example of

   a.   a randomized comparative experiment.

   b.   an experiment which is not double-blind.

   c.   bad sampling methods.

8. To determine interest in a new paper towel, samples are mailed to 300 local residents, of which 120 prefer it to their current brand. What is the standard deviation of the sampling distribution of this statistic?

   a.   2.83%

   b.   3.65%

   c.   4.47%

9. A random poll of 600 people shows that 60% of those polled are in favor of a new school building. Find a 95% confidence interval for the proportion of the residents in favor of a new school building.

   a.   56% to 64%

   b.   40% to 80%

   c.   58% to 62%

10. An election poll of 628 voters showed that 412 of them approve of the policies of the leader. A 95% confidence interval for the proportion of voters who approve of the policies of the leader is

    _____.

    a.   61.8% to 69.4%

    b.   63.7% to 67.5%

    c.   65.4% to 65.8%

Word Search

Refer to pages 281 – 282 of your text to obtain the Review Vocabulary. There are 20 hidden vocabulary words/expressions in the word search below. *Randomized comparative experiment, Nonresponse, 95% confidence interval, Undercoverage,* and *Statistical inference* do not appear in the word search due to expression length. *Statistical significance, statistic, sample, sample proportion, experiment* and *double-blind experiment* appear separately in the word search. It should be noted that spaces and hyphens are removed.

```
O R T E R T E P I P A Q A P A I E D Y T O E F Z F
G S C E O B S E R V A T I O N A L S T U D Y T O W
N T E P E L P M A S E C N E I N E V N O C N J M B
T H F L E C A N N E L P M A S O R L E H E X P C T
N B F J X E R O R R E F O N I G R A M M T O H O A
E X E L P M A S M O D N A R E L P M I S B S O N A
M R O H F S M P S F V N E H I E E R R M N B J T M
I T B P R T E S P X H O N Q F T E S E W K I G R I
R G E T S I T E H S A M P L E P R O P O R T I O N
E S C E P Z E N S M N O L S X E E R X W N O S L O
P X A O V O R L H S I I M E O S E C E D S Z E G I
X I L A Y P E D R F O E G G V D E G D T Z W I R T
E L P M A S E S N O P S E R Y R A T N U L O V O A
E C N A C I F I N G I S L A C I T S I T A T S U L
M I P O I D E A D T J L A Z K K R H L B Y S A P U
E R X O L O G U Y G F M P F F X A T B L I X R E P
D Y O F I O B L H S E X S D Y A D R E A V A B E O
R C X N O I T U B I R T S I D G N I L P M A S H P
S S S T I G I D M O D N A R F O E L B A T R J L D
R W N P A I G A S N H Z O I M F G F U E G W R C C
T O O E B Y D U T S E V I T C E P S O R P U H D F
C F D A S C T V L C S C O N F O U N D I N G F M F
P O I V L G A G O A D X E I E D H T R E N I K J H
E E N S G R W X E Y S T P E Y C M M F N C A G E N
I F S T A T I S T I C E O E N E E Q E B G Y I C A
```

1. _____   11. _____

2. _____   12. _____

3. _____   13. _____

4. _____   14. _____

5. _____   15. _____

6. _____   16. _____

7. _____   17. _____

8. _____   18. _____

9. _____   19. _____

10. _____   20. _____

Chapter 8
Probability: The Mathematics of Chance

Chapter Objectives

Check off these skills when you feel that you have mastered them.

- [ ] Explain what is meant by random phenomenon.

- [ ] Describe the sample space for a given random phenomenon.

- [ ] Explain what is meant by the probability of an outcome.

- [ ] Describe a given probability model by its two parts.

- [ ] List and apply the four rules of probability and be able to determine the validity/invalidity of a probability model by identifying which rule(s) is (are) not satisfied.

- [ ] Compute the probability of an event when the probability model of the experiment is given.

- [ ] Apply the addition rule to calculate the probability of a combination of several disjoint events.

- [ ] Draw the probability histogram of a probability model, and use it to determine probabilities of events.

- [ ] Explain the difference between a discrete and a continuous probability model.

- [ ] Determine probabilities with equally likely outcomes.

- [ ] Use the fundamental principal of counting to determine the number of possible outcomes involved in an event and/or the sample space.

- [ ] List two properties of a density curve.

- [ ] Construct basic density curves that involve geometric shapes (rectangles and triangles) and utilize them in determining probabilities.

- [ ] State the mean and calculate the standard deviation of a sample statistic $\left(\hat{p}\right)$ taken from a normally distributed population.

- [ ] Explain and apply the 68–95–99.7 rule to compute probabilities for the value of $\hat{p}$ from a single simple random sample (SRS).

- [ ] Compute the mean (μ) and standard deviation (σ) of an outcome when the associated probability model is defined.

- [ ] Explain the significance of the law of large numbers.

- [ ] Explain the significance of the central limit theorem.

Guided Reading

Introduction

Games of chance are an application of the laws of randomness, but much more fundamental areas of human and natural activity are subject to these laws. Physics, genetics, economics, politics, and essentially any area in which large numbers of people or objects are examined or measured can best be understood via the mathematics of chance.

⚷ Key idea

Like a roll of the dice or a coin flip, a repeatable phenomenon is **random** if any particular outcome is quite unpredictable, while in the long run, after a large number of repeated trials, a regular, predictable pattern emerges.

Section 8.1   Probability Models and Rules

If you perform an experiment of tossing a coin, throwing a die, or choosing a simple random sample (SRS), the outcome will not be known in advance. However, after many repetitions, a regular pattern will emerge.

⚷ Key idea

The **probability** of any outcome of a random phenomenon is the proportion of times the outcome would occur in a very long series of repetitions.

⚷ Key idea

The **sample space**, S, of a random phenomenon is the set of all possible outcomes.

⚷ Key idea

An **event** is any outcome or any set of outcomes of a random phenomenon. That is, an event is a subset of the sample space.

⚷ Key idea

A **probability model** is a mathematical description of a random phenomenon consisting of two parts as follows.
- a sample space, S
- a way of assigning probabilities to events

⚷ Key idea

If A and B are events in sample space S, and $P(A)$ is the probability of that event, then the following hold true.
- $0 \le P(A) \le 1$: Any probability is a number between 0 and 1, inclusively.
- $P(S) = 1$: All possible outcomes together must have probability 1.
- $P(A \text{ or } B) = P(A) + P(B)$: If two events ($A$ and B) have no outcomes in common, the probability that one or the other occurs is the sum of their individual probabilities. This is the **addition rule for disjoint events**.
- $P(A^c) = 1 - P(A)$: The probability that an event does not occur is 1 minus the probability that the event does occur. A^c is the complement of event A, which is in sample space S.

⟋ Example A

Consider tossing a die and flipping a coin.
a)   Determine the sample space.
b)   Assume a value of "0" was assigned to heads and "1" for tails. Sum together the value on the die with the assigned value on the coin. What is the probability model?

Solution

a)   The sample space is {1H, 2H, 3H, 4H, 5H, 6H, 1T, 2T, 3T, 4T, 5T, 6T}.

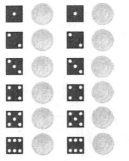

b)   The probability model is as follows.

Outcome	1	2	3	4	5	6	7
Probability	$\frac{1}{12}$	$\frac{2}{12}=\frac{1}{6}$	$\frac{2}{12}=\frac{1}{6}$	$\frac{2}{12}=\frac{1}{6}$	$\frac{2}{12}=\frac{1}{6}$	$\frac{2}{12}=\frac{1}{6}$	$\frac{1}{12}$

	Sum			Sum
0	1		1	2
0	2		1	3
0	3		1	4
0	4		1	5
0	5		1	6
0	6		1	7

✎ Question 1

Consider tossing a die and flipping a coin.
a)   Assume a value of "1" was assigned to heads and "3" for tails. Sum together the value on the die with the assigned value on the coin. What is the probability model?
b)   What is the probability that the sum is an even number?
c)   What is the complement to the event: sum is 2?
d)   What is the probability that the sum is not 4?

Answer

a)   The probability model is as follows.

Outcome	2	3	4	5	6	7	8	9
Probability	$\frac{1}{12}$	$\frac{1}{12}$	$\frac{2}{12}=\frac{1}{6}$	$\frac{2}{12}=\frac{1}{6}$	$\frac{2}{12}=\frac{1}{6}$	$\frac{2}{12}=\frac{1}{6}$	$\frac{1}{12}$	$\frac{1}{12}$

b) $\frac{1}{2}$
c)   the sum is not 2
d) $\frac{5}{6}$

⌐ Key idea

The **probability histogram** of a probability model shows graphically the likelihood of each outcome. The height of each bar shows the probability of the outcome at its base, and the sum of the heights is 1.

⌐ Example B

Construct the probability histogram for the probability model in Example A.

Solution

The events or obtaining a sum of 1 or 7 each have probability $\frac{1}{12}$. The other 5 events each have a probability of $\frac{1}{6}$. The probabilities are indicated by the height of each rectangle.

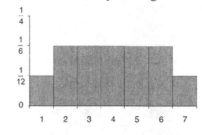

Section 8.2   Discrete Probability Models

⌐ Key idea

A probability model with a finite sample space is called **discrete**. In a discrete probability model, you are able to individually list out all events in the sample space and assign probabilities to each event. These probabilities must be numbers between 0 and 1 and must have sum 1.

⌐ Key idea

The probability of a collection of outcomes (an event) is the sum of the probabilities of the outcomes that constitute the event.

⌐ Example C

If you roll two dice, what is the probability of rolling a sum less than or equal to 6?

Solution

A total of 6 or less means one of these events: {2, 3, 4, 5, 6}. Add up their probabilities, which can be read from the following histogram.

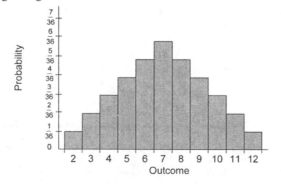

Since $P(2)+P(3)+P(4)+P(5)+P(6)=\frac{1}{36}+\frac{2}{36}+\frac{3}{36}+\frac{4}{36}+\frac{5}{36}=\frac{15}{36}=\frac{5}{12}$, we have the following.

$$P(\text{sum less than } 6)=\frac{5}{12}$$

✏️ **Question 2**

Consider the situation in Question 1.

a) What is the probability that the sum is more than 3?
b) What is the probability that the sum is less than 3?
c) Should the answers to Parts a and b sum to be 1? Explain.

Answer

a) $\dfrac{5}{6}$

b) $\dfrac{1}{12}$

c) no

Section 8.3   Equally Likely Outcomes

🔑 **Key idea**

If a random phenomenon has k possible outcomes, all **equally likely**, then each individual outcome has probability $\frac{1}{k}$. The probability of any event A is as follows.

$$P(A) = \frac{\text{count of outcomes in } A}{\text{count of outcomes in } S} = \frac{\text{count of outcomes in } A}{k}$$

〰️ **Example D**

When rolling two ordinary dice, there are 36 possible outcomes: $(1,1)$, $(1,2)$, and so on up to $(6,6)$.

a) What is the probability of an outcome of $(1,1)$?
b) What is the probability that one of the die is a 1 and the other is a 2?

Solution

a) Since each of the outcomes is equally likely, each has the same probability. Thus, $P\left[(1,1)\right] = \frac{1}{36}$.

b) Since it is possible to roll $(1,2)$ or $(2,1)$, $P\left[(1,2) \text{ or } (2,1)\right] = \frac{2}{36} = \frac{1}{18}$.

✏️ **Question 3**

When rolling two dice, what is the probability of obtaining a sum of 10?

Answer

$\dfrac{1}{12}$

🔑 **Key idea**

With equally likely outcomes, probability calculations come from **combinatorics**, or the study of counting methods.

- Rule A: Arranging k objects chosen from a set of n possibilities, with **repetitions allowed**, can be done in $n \times n \times ... \times n = n^k$ distinct ways.
- Rule B: Arranging k objects chosen from a set of n possibilities, with **no repetitions allowed**, can be done in $n \times (n-1) \times ... \times (n-k+1)$. (notice there are k factors here).

⤳ Example E

a)   How many code words (that is, strings of letters) of length four can be formed that use only the vowels {A, E, I, O, U, Y}?

b)   How many of these words have no letter occurring more than once?

Solution

a)   Since repetition is not excluded, we will use Rule A where $n = 6$ and $k = 4$.
$$6 \times 6 \times 6 \times 6 = 6^4 = 1296$$

b)   Since repetition is not allowed, we will use Rule B where $n = 6$ and $k = 4$.
$$6 \times (6-1) \times ... \times (6-4+1) = 6 \times 5 \times 4 \times 3 = 360$$

✎ Question 4

Consider an identification code, which made up of three letters of the alphabet followed by three digits. What is the probability that a randomly chosen identification code is ABC123 given that

a)   repetition is allowed?

b)   repetition is not allowed?

Answer

a) $\dfrac{1}{17,576,000}$

b) $\dfrac{1}{11,232,000}$

Section 8.4   Continuous Probability Models

⌗ Key idea

In a **continuous probability model**, there are infinitely many possible events that could occur. In order to assign probabilities to events, we look at area under a *density curve*. The total area under the curve bounded by a horizontal axis must equal 1. The probability of a single value to occur is 0. In the case of continuous probability models, you will be looking at the probability of a range of values (an interval) to occur.

⌗ Key idea

The **uniform probability model** has a density curve that creates a rectangle along a horizontal axis. The area of the rectangle will always be 1 for the continuous uniform model.

⤳ Example F

Suppose you specify that the range of a random number generator is to be all numbers between 0 and 4. The density curve for the outcome has constant height between 0 and 4, and height 0 elsewhere.

a)   Draw a graph of the density curve.

b)   Suppose the generator produces a number X. Find $P(X \leq 2.7)$.

c)   Find $P(1.2 \leq X \leq 3.6)$.

Solution

a)   Since the width of the base is 4, the height would be $\frac{1}{4} = 0.25$.

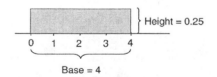

b)   Since the area of a rectangle base $\times$ height, the probability will be $(2.7)(0.25) = 0.675$.

c)   We need to determine the length of the base. It will be $3.6 - 1.2 = 2.4$. Thus, the probability will be Area = base $\times$ height $= (2.4)(0.25) = 0.6$.

✏ **Question 5**

Generate two random numbers between 0 and 3 and take their sum. The sum can take any value between 0 and 6. The density curve is the triangle.
a)   What is the probability that the sum is less than 2?
b)   What is the probability that the sum is between 1.5 and 4?

Answer

a) $\dfrac{2}{9}$
                                                                b) $\dfrac{47}{72}$

⊶ **Key idea**

Normal distributions are continuous probability models. The 68–95–99.7 rule applies to a normal distribution and we can use it for determining probabilities. It is useful in determining the proportion of a population with values falling in certain ranges.

⊶ **Key idea**

The sample proportion, $\hat{p}$, will vary from sample to sample according to a normal distribution with mean, p, and standard deviation $\sqrt{\frac{p(1-p)}{n}}$, where n is the number in the sample. (p is the population proportion.)

〜 **Example F**

Suppose 62% of all children under the age of 6 watch a certain TV show, say the *Captain Buckaroo Show*. You choose 210 children under the age of 6 to sample at random. What are the mean and standard deviation of the proportion of children under the age of 6 that watch the *Captain Buckaroo Show*?

Solution

The population proportion of children under the age of 6 that watch the *Captain Buckaroo Show* is $p = 0.62$. The sample proportion, $\hat{p}$, of children under the age of 6 that watch the *Captain Buckaroo Show* in a random sample of $n = 210$ has mean $p = 0.62$ and standard deviation as follows.

$$\sqrt{\frac{p(1-p)}{n}} = \sqrt{\frac{0.62(1-0.62)}{210}} \sqrt{\frac{0.62(0.38)}{210}} = \sqrt{\frac{0.2356}{210}} \approx 0.033$$

✏ **Question 6**

Suppose 12% of all adults over the age of 25 watch a certain TV show, say the *Captain Buckaroo Show*. You choose 171 adults over the age of 25 to sample at random. By applying the 68–95–99.7 rule, determine 95% of the time the sample proportion will be in what interval? Round your answer to the nearest tenth of a percent.

Answer
7.0% to 17.0%

Section 8.5   The Mean and Standard Deviation of a Probability Model

⊶ **Key idea**

The **mean** of a discrete probability model is the sum of the possible outcomes times the probability of each outcome. If there are k possible outcomes in the sample space, then there will be k terms in the sum. Each term will have a probability associated with it. The sum of all the probabilities will be 1.

⌔ Example G
What is the mean sum in Example C?

Solution
There are 11 possible outcomes. Thus, there are 11 terms to sum.

$$\frac{1}{36}(2)+\frac{2}{36}(3)+\frac{3}{36}(4)+\frac{4}{36}(5)+\frac{5}{36}(6)+\frac{6}{36}(7)+\frac{5}{36}(8)+\frac{4}{36}(9)+\frac{3}{36}(10)+\frac{2}{36}(11)+\frac{1}{36}(12)=$$

$$\frac{2}{36}+\frac{6}{36}+\frac{12}{36}+\frac{20}{36}+\frac{30}{36}+\frac{42}{36}+\frac{40}{36}+\frac{36}{36}+\frac{30}{36}+\frac{22}{36}+\frac{12}{36}=\frac{252}{36}=7$$

Due to the symmetry of the probability histogram, this mean should be intuitively obvious.

✐ Question 7
Consider the situation in Question 1. What is the mean sum?

Answer
5.5

⚷ Key idea
The **law of large numbers** states that as a random phenomenon is repeated a large number of times, the mean of the trials, $\bar{x}$, gets closer and closer to the mean of the probability model, μ.

⚷ Key idea
The **variance** of a discrete probability model that has numerical outcomes $x_1, x_2, ..., x_k$ in a sample space will have variance as follows.

$$\sigma^2 =\left(x_1-\mu\right)^2 p_1 +\left(x_2-\mu\right)^2 p_2 +...+\left(x_k-\mu\right)^2 p_k,$$

where p_j is the probability of outcome x_j. The **standard deviation** σ is the square root of the variance.

⌔ Example H
What is the standard deviation in Example C?

Solution
There are 11 possible outcomes. Thus, there are 11 terms to sum.

$$\sigma^2 =(2-7)^2\left(\tfrac{1}{36}\right)+(3-7)^2\left(\tfrac{2}{36}\right)+(4-7)^2\left(\tfrac{3}{36}\right)+(5-7)^2\left(\tfrac{4}{36}\right)+(6-7)^2\left(\tfrac{5}{36}\right)$$
$$+(7-7)^2\left(\tfrac{6}{36}\right)+(8-7)^2\left(\tfrac{5}{36}\right)+(9-7)^2\left(\tfrac{4}{36}\right)+(10-7)^2\left(\tfrac{3}{36}\right)$$
$$+(11-7)^2\left(\tfrac{2}{36}\right)+(12-7)^2\left(\tfrac{1}{36}\right)$$
$$=(-5)^2\left(\tfrac{1}{36}\right)+(-4)^2\left(\tfrac{2}{36}\right)+(-3)^2\left(\tfrac{3}{36}\right)+(-2)^2\left(\tfrac{4}{36}\right)+(-1)^2\left(\tfrac{5}{36}\right)+0^2\left(\tfrac{6}{36}\right)$$
$$+1^2\left(\tfrac{5}{36}\right)+2^2\left(\tfrac{4}{36}\right)+3^2\left(\tfrac{3}{36}\right)+4^2\left(\tfrac{2}{36}\right)+5^2\left(\tfrac{1}{36}\right)$$
$$=25\left(\tfrac{1}{36}\right)+16\left(\tfrac{2}{36}\right)+9\left(\tfrac{3}{36}\right)+4\left(\tfrac{4}{36}\right)+1\left(\tfrac{5}{36}\right)+0\left(\tfrac{6}{36}\right)+1\left(\tfrac{5}{36}\right)+4\left(\tfrac{4}{36}\right)+9\left(\tfrac{3}{36}\right)+16\left(\tfrac{2}{36}\right)+25\left(\tfrac{1}{36}\right)$$
$$=\tfrac{25}{36}+\tfrac{32}{36}+\tfrac{27}{36}+\tfrac{16}{36}+\tfrac{5}{36}+\tfrac{0}{36}+\tfrac{5}{36}+\tfrac{16}{36}+\tfrac{27}{36}+\tfrac{32}{36}+\tfrac{25}{36}=\tfrac{210}{36}$$

Thus, the standard deviation is $\sigma=\sqrt{\tfrac{210}{36}}\approx 2.4152$.

✐ Question 8
Consider the situation in Question 1. What is the standard deviation of the sum?

Answer
1.9791

Section 8.6   The Central Limit Theorem

⌐ Key idea

The **central limit theorem** says that the distribution of any random phenomenon tends to be normal if we average it over a large number of independent repetitions. It also says that a sample distribution will have the same mean, μ, as the original phenomenon. It will also have a standard deviation

equal to $\dfrac{\sigma}{\sqrt{n}}$, where σ is the standard deviation of a single trial and n is the number of trials.

⌐ Example I

Suppose a marketing exam had scores which were normally distributed with a mean of 73 and a standard deviation of 12.

a)   Suppose you chose 10 of the students at random and computed the mean, $\bar{x}$, of their scores. What is the standard deviation of $\bar{x}$?

b)   How large a sample size n would you have to use to bring the standard deviation of $\bar{x}$ down to 2?

c)   By taking a large number of samples of 10 students chosen at random, what range of exam scores will contain the middle 95% of the many $\bar{x}$'s?

Solution

a)   Take the standard deviation of the original distribution, which is 12, and divide by the square root of 10; we get $\frac{12}{\sqrt{10}} \approx 3.7947$.

b)   To bring it down from 12 to 2 or less, we would have to divide by 6 or more. If the square root of the sample size n is 6 or greater, then n must be at least 36.

c)   Sample means $\bar{x}$ have a sampling distribution close to normal with mean $\mu = 73$ and standard deviation approximately 3.7947.   Therefore, 95% of all samples have an $\bar{x}$ between $73 - 2(3.7947) = 73 - 7.5894 = 65.4106$ and $73 + 2(3.7947) = 73 + 7.5894 = 80.5894$.   With rounding, we would say between 65 and 81.

⌐ Example J

Consider the following spinner game. It costs $1 to play. If you spin a negative value, you lose your dollar as well as the additional amount indicated ($1 or $3). If you spin a positive value, you keep your dollar and you receive the additional amount indicated ($5 or $3).

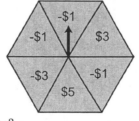

a)   What is the mean of a single game?

b)   What is the standard deviation?

c)   What is the mean and the standard deviation of the average win/loss of the game if it was played 100 times in one day?

d)   Apply the 99.7 part of the 68-95-99.7 rule to determine a range of average win/loss of playing this game 100 times per day for 1000 days.

Solution

The probability model would be as follows.

Outcome	−$2	−$2	−$4	$5	−$2	$3
Probability	$\frac{1}{6}$	$\frac{1}{6}$	$\frac{1}{6}$	$\frac{1}{6}$	$\frac{1}{6}$	$\frac{1}{6}$

a) The mean of a single game would be as follows.

$$\mu = (-2)\left(\tfrac{1}{6}\right)+(-2)\left(\tfrac{1}{6}\right)+(-4)\left(\tfrac{1}{6}\right)+(5)\left(\tfrac{1}{6}\right)+(-2)\left(\tfrac{1}{6}\right)+(3)\left(\tfrac{1}{6}\right)$$

$$= \tfrac{-2}{6}+\tfrac{-2}{6}+\tfrac{-4}{6}+\tfrac{5}{6}+\tfrac{-2}{6}+\tfrac{3}{6} = \tfrac{-2}{6} = -\tfrac{1}{3} \approx -\$0.33$$

b) The variance of a single game would be as follows.

$$\sigma^2 = \left[-2-\left(-\tfrac{1}{3}\right)\right]^2\left(\tfrac{1}{6}\right)+\left[-2-\left(-\tfrac{1}{3}\right)\right]^2\left(\tfrac{1}{6}\right)+\left[-4-\left(-\tfrac{1}{3}\right)\right]^2\left(\tfrac{1}{6}\right)+\left[5-\left(-\tfrac{1}{3}\right)\right]^2\left(\tfrac{1}{6}\right)$$

$$+\left[-2-\left(-\tfrac{1}{3}\right)\right]^2\left(\tfrac{1}{6}\right)+\left[3-\left(-\tfrac{1}{3}\right)\right]^2\left(\tfrac{1}{6}\right)$$

$$=\left(-\tfrac{5}{3}\right)^2\left(\tfrac{1}{6}\right)+\left(-\tfrac{5}{3}\right)^2\left(\tfrac{1}{6}\right)+\left(-\tfrac{11}{3}\right)^2\left(\tfrac{1}{6}\right)+\left(\tfrac{16}{3}\right)^2\left(\tfrac{1}{6}\right)+\left(-\tfrac{5}{3}\right)^2\left(\tfrac{1}{6}\right)+\left(\tfrac{10}{3}\right)^2\left(\tfrac{1}{6}\right)$$

$$=\left(\tfrac{25}{9}\right)\left(\tfrac{1}{6}\right)+\left(\tfrac{25}{9}\right)\left(\tfrac{1}{6}\right)+\left(\tfrac{121}{9}\right)\left(\tfrac{1}{6}\right)+\left(\tfrac{256}{9}\right)\left(\tfrac{1}{6}\right)+\left(\tfrac{25}{9}\right)\left(\tfrac{1}{6}\right)+\left(\tfrac{100}{9}\right)\left(\tfrac{1}{6}\right)$$

$$=\tfrac{25}{54}+\tfrac{25}{54}+\tfrac{121}{54}+\tfrac{256}{54}+\tfrac{25}{54}+\tfrac{100}{54}=\tfrac{552}{54}=\tfrac{92}{9}$$

Thus, the standard deviation would be $\sqrt{\tfrac{92}{9}} \approx 3.1972$.

c) From the central limit theorem, the mean would be approximately −$0.33. The standard deviation would be $\tfrac{3.1972}{\sqrt{100}} \approx 0.3197$.

d) Almost all, 99.7%, of all daily win/losses would fall within three standard deviations of the mean. Thus, the total win/losses after playing 100 times will fall between $-0.33-3(0.3197)=-0.33-0.9591=-0.6291$ and $-0.33+3(0.3197)=-0.33+0.9591=1.2891$.

With rounding, we would say between −$0.63 and $1.29.

✎ Question 9

Consider the following spinner game. It costs $1 to play. If you spin a negative value, you lose your dollar as well as the additional amount indicated ($1, $2, or $3). If you spin a positive value, you keep your dollar and you receive the additional amount indicated ($4 or $7).

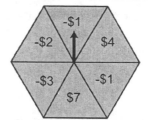

a) What is the mean and standard deviation of a single game?

b) What is the mean and the standard deviation of the average win/loss of the game if it was played 1000 times in one day?

c) Apply the 99.7 part of the 68-95-99.7 rule to determine a range of average win/loss of playing this game 1000 times per day for 1000 days.

Answer

a) $\mu = \$0$ and $\sigma \approx 4.0415$

b) The mean would be $0 and the standard deviation would be approximately 0.1278

c) between −$0.38 and $0.38

Homework Help

☞ To assist you in homework, a page of "blank" normal distributions appears after this section.

Exercises 1 – 2
Since these exercises involve actual experiments, results will vary.

Exercise 3
Count up the number of zeros and determine the proportion of zeros in the first 200 digits. The partial table below should be helpful.

TABLE 7.1	Random	Digits					
101	19223	95034	05756	28713	96409	12531	42544	82853
102	73676	47150	99400	01927	27754	42648	82425	36290
103	45467	71709	77558	00095	32863	29485	82226	90056
104	52711	38889	93074	60227	40011	85848	48767	52573
105	95592	94007	69971	91481	60779	53791	17297	59335
106	68417	35013	15529	72765	85089	57067	50211	47487

Exercise 4
Probability is a number between 0 and 1, inclusive. The higher the number, the higher the probability that the event will occur. The lower the number, the lower the probability that the event will occur.

Exercise 5
(a) There are 11 elements in this sample space.
(b) There are 11 elements in this sample space.
(c) There are 2 elements in this sample space.

Exercise 6
(a) There are 2 elements in this sample space.
(b) There are 15 elements in this sample space.
(c) Answers will vary. Use judgment for lower and upper limits.

Exercise 7
(a) There are 16 elements in this sample space. Be systematic when listing elements of the sample space.
(b) There are 5 elements in this sample space.

Exercise 8
(a) There are 2 elements in this sample space.
(b) Answers will vary. Use judgment for lower and upper limits.
(c) Answers will vary. Use judgment for lower and upper limits.

Exercise 9
(a) Sum the probabilities in the table together. Use Probability Rule 4 from Section 8.1.
(b) It is assumed in this exercise that adult and scam are disjoint events. Use Probability Rule 3 from Section 8.1.

Exercise 10
(a) Use Probability Rule 4 from Section 8.1.
(b) Use Probability Rule 4 from Section 8.1 or sum the probabilities of three disjoint events together (this relies on part a).

Exercise 11
Answers will vary. Any two events that can occur together will do.

Exercise 12
(a) Sum the probabilities in the table together. Use Probability Rule 4 from Section 8.1.
(b) Use Probability Rule 3 from Section 8.1.

Exercise 13
(a) The following may be helpful in creating a probability histogram.

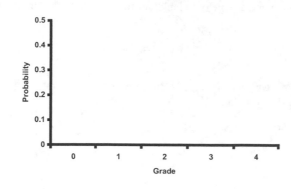

(b) You will need to sum together two probabilities.

Exercise 14
The following may be helpful in creating probability histograms.

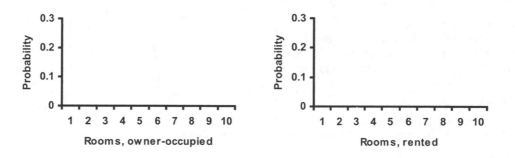

Exercise 15
For each a part make sure the probabilities are between 0 and 1, inclusively, and have sum 1.

Exercise 16
For both owner-occupied units and rented units, find $P(5,6,7,8,9,10)$ by summing the appropriate probabilities.

Exercise 17
As recommended in the exercise, make a drawing showing all possibilities. There are 36 total.
Determine how many times each sum occurs and fill in the following table.

Outcome	1	2	3	4	5	6	7	8	9	10	11	12
Probability												

Exercise 18
As recommended in the exercise, make a drawing showing all possibilities. There are 16 total.
Determine how many times each sum occurs and fill in the following table.

Intelligence	3	4	5	6	7	8	9
Probability							

Determine the probability of intelligence 7 or higher by summing the appropriate 3 probabilities.

Exercise 19
Realizing that all 90 guests are equally likely to get the prize, determine $P(\text{woman})$.

Exercise 20
(a) Use the first letters to stand for names and write the 10 possible choices.
(b) – (d) Count up the number of possible choices that satisfy each condition in order to determine the
      probabilities.

Exercises 21 – 25
Carefully read Section 8.3 before responding to these exercises. You will need to utilize the rules of
counting arrangements of distinct items. In Exercise 25, first determine the 6 possible arrangements
of the letters.

Exercises 26 – 27
Carefully read Section 8.3 before responding to these exercises. You will need to utilize the rules of
counting arrangements of distinct items. In each exercise you will need to sum together the possible
outcomes of the individual types in order answer each question. You will then need to apply the
formula in the definition of equally likely outcomes.

Exercise 28
(a) Determine the probability for each square face. Use the fact that a face is either a square or a
      triangle to determine the probability of obtaining a triangle. Finally, determine the probability
      for each triangle.
(b) Answers will vary. Start with a different probability for squares and follow the same procedure
      as in part a.

Exercise 29
Draw the density curve and shade the appropriate region for each part. You will need that the area of
a triangle is $\frac{1}{2} \times \text{base} \times \text{height}$.

Exercise 30
Draw the rectangular density curve and shade the appropriate region for each part. Determine the
height of the rectangular density curve by realizing that base is of length 2 and area of a rectangle is
base $\times$ height. The area under the density curve will equal 1.

Exercise 31
The following may be helpful in this exercise.

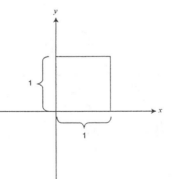

Exercise 32
Carefully read Section 8.5 before responding to this exercise. Realizing that earnings are $400 times sales, fill in the following table.

Earnings				
Probability	0.3	0.4	0.2	0.1

Use the definition of the mean of a discrete probability model. You will need to sum four terms.

Exercise 33
Carefully read Section 8.5 before responding to this exercise. Use the definitions of the mean of a discrete probability model and the standard deviation of a discrete probability model. For each you will need to sum five terms. To find the standard deviation, you will need to take the square root of the variance. Try to make your intermediate calculations as accurate as possible to avoid round-off error.

Exercises 34 – 36
Carefully read Section 8.5 before responding to these exercises. Use the definition of the mean of a discrete probability model. Try to make your intermediate calculations as accurate as possible to avoid round-off error. In Exercise 35, recreate (or use) the probability histograms created in Exercise 14 and locate the mean on each histogram.

Exercise 37
Note the symmetry in both density curves.

Exercise 38
Answers will vary. Use the law of large numbers in your explanation.

Exercise 39
Carefully read Section 8.5 before responding to these exercise. Fill in the following table for Part b.

Outcome	2	3	4	5	6	7	8	9	10	11	12
Probability											

Use the definition of the mean of a discrete probability model. Try to make your intermediate calculations as accurate as possible to avoid round-off error. In Part c, answers will vary. Remember though, expected values are averages, so they behave like averages.

Exercise 39
Carefully read Section 8.5 before responding to this exercise. Fill in the following table for Part a.

Outcome	Win $2	Lose $1
Probability		

In Part b, use the definitions of the mean of a discrete probability model and the standard deviation of a discrete probability model. For each you will need to sum five terms. To find the standard deviation, you will need to take the square root of the variance. Try to make your intermediate calculations as accurate as possible to avoid round-off error. Use the results of Part b (mean) and the law of large numbers to answer Part c.

Exercises 41 – 42
Carefully read Section 8.5 before responding to these exercise. Use the definition of the mean of a discrete probability model. Try to make your intermediate calculations as accurate as possible to avoid round-off error.

Exercise 43

Carefully read Section 8.5 before responding to this exercise.

(a) Sum the probabilities in the table together. Use Probability Rule 4 from Section 8.1.

(b) Fill in the following table. The first five outcomes will be negative. The last outcome will be positive. The final probability is from part a.

Probability	Outcome
0.00039	
0.00044	
0.00051	
0.00057	
0.00060	

Use the table and the definition of the mean of a discrete probability model. Try to make your intermediate calculations as accurate as possible to avoid round-off error.

Exercise 44

As instructed, use the definitions of the mean and variance of a discrete probability model. Remember that $\mu - \sigma$ and $\mu + \sigma$ are outcomes.

Exercise 45

Carefully read Section 8.6 before responding to this exercise. Sample means $\overline{x}$ have a sampling distribution close to normal with mean $\mu = 0.15$. Calculate the standard deviation $\dfrac{\sigma}{\sqrt{n}}$ and use the fact that 95% of all samples have an $\overline{x}$ between $\mu - 2\left(\dfrac{\sigma}{\sqrt{n}}\right)$ and $\mu + 2\left(\dfrac{\sigma}{\sqrt{n}}\right)$.

Exercise 46

(a) Use the symmetry of the normal distribution and the 68 part of the 68-95-99.7 rule.

(b) Sample means $\overline{x}$ have a sampling distribution close to normal with mean $\mu = 300$. Calculate the standard deviation $\dfrac{\sigma}{\sqrt{n}}$. Use the symmetry of the normal distribution and the 95 part of the 68-95-99.7 rule.

Exercise 47

(a) Calculate the standard deviation $\dfrac{\sigma}{\sqrt{n}}$.

(b) Since we want to cut the standard deviation in half (from 10 mg to 5 mg), determine what value of n will make $\dfrac{\sigma}{2} = \dfrac{\sigma}{\sqrt{n}}$. Additional answers will vary.

Exercise 48

The average winnings per bet has the mean, μ, from Exercise 40 for any number of bets. The standard deviation of the average winnings is $\dfrac{\sigma}{\sqrt{n}}$ (where σ is also from Exercise 40). For both Parts a and b, calculate $\dfrac{\sigma}{\sqrt{n}}$ with the values of n. Then for each part, determine the spread of average winnings $\mu - 3\left(\dfrac{\sigma}{\sqrt{n}}\right)$ to $\mu + 3\left(\dfrac{\sigma}{\sqrt{n}}\right)$.

Exercise 49

(a) For $n = 1$, sketch a normal curve and mark the center. The change-of-curvature points are one standard deviation from the center. Extend the curve three standard deviations in both directions. On the same graph, do the same for $n = 3$

(b) Use the 95 part of the 68-95-99.7 rule with $\sigma = 10$.

(c) Use the 95 part of the 68-95-99.7 rule with $\sigma = 5.77$.

Exercise 50

Carefully read Section 8.6 before responding to this exercise. Use the definition of the standard deviation of a discrete probability model. You will need to sum seven terms. To find the standard deviation, you will need to take the square root of the variance. Try to make your intermediate calculations as accurate as possible to avoid round-off error. Apply the central limit theorem. Sample means $\overline{x}$ have a sampling distribution close to normal with mean $\mu = 6$. Calculate the standard deviation $\dfrac{\sigma}{\sqrt{n}}$ and use the fact that 68% of all samples have an $\overline{x}$ between $\mu - \dfrac{\sigma}{\sqrt{n}}$ and $\mu + \dfrac{\sigma}{\sqrt{n}}$.

Exercise 51

Determine the ACT exam scores that are 1, 2, and 3 standard deviations from the mean and label them on the graph below.

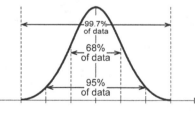

ACT exam scores

For Part b, use the fact that sample means $\overline{x}$ have a sampling distribution close to normal with mean $\mu = 20.8$. Calculate the standard deviation $\dfrac{\sigma}{\sqrt{n}}$. In Part c, use the symmetry of the normal curve along with the 68-95-99.7 rule.

Exercise 52

In Part a, the sample proportion, $\hat{p}$, has mean p and standard deviation $\sqrt{\dfrac{p(1-p)}{n}}$. In Part b, use the symmetry of the normal curve along with the 68-95-99.7 rule.

Exercises 53 – 54
Carefully read Section 8.3 before responding to these exercises. You will need to utilize the rules of counting arrangements of distinct items.

Exercise 55
(a) Use Probability Rule 3 from Section 8.1.
(b) Use Probability Rule 4 from Section 8.1.

Exercise 56
Use the definition of the standard deviation of a discrete probability model. You will need to sum eight terms. Read and apply the law of large numbers (Section 8.5).

Exercise 57
(a) Carefully read Section 8.5 before responding to this exercise. Use the definition of the standard deviation of a discrete probability model to first find the variance with $\mu = 1.03$. You will need to sum eight terms. To find the standard deviation, you will need to take the square root of the variance. Try to make your intermediate calculations as accurate as possible to avoid round-off error.

(b) The mean, $\bar{x}$, has mean $\mu = 1.03$ and standard deviation $\dfrac{\sigma}{\sqrt{n}}$. The central limit theorem says

that $\bar{x}$ is approximately normal with this mean and standard deviation. Use the 95 part of the 68-95-99.7 rule.

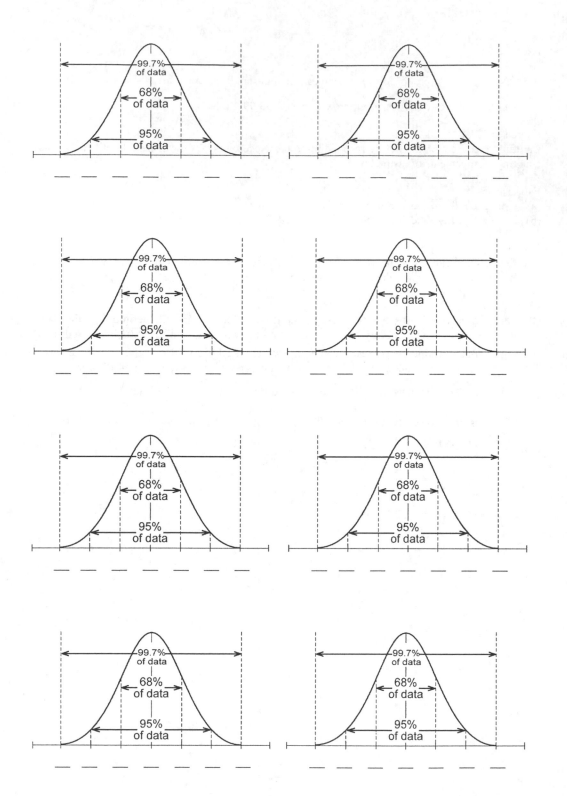

Do You Know the Terms?

Cut out the following 18 flashcards to test yourself on Review Vocabulary. You can also find these
flashcards at http://www.whfreeman.com/fapp7e.

Chapter 8 Probability: The Mathematics of Chance **Addition rule**	Chapter 8 Probability: The Mathematics of Chance **Central limit theorem**
Chapter 8 Probability: The Mathematics of Chance **Combinatorics**	Chapter 8 Probability: The Mathematics of Chance **Complement rule**
Chapter 8 Probability: The Mathematics of Chance **Continuous probability model**	Chapter 8 Probability: The Mathematics of Chance **Density curve**
Chapter 8 Probability: The Mathematics of Chance **Discrete probability model**	Chapter 8 Probability: The Mathematics of Chance **Disjoint events**

The average of many independent random outcomes is approximately normally distributed. When we average n independent repetitions of the same random phenomenon, the resulting distribution of outcomes has mean equal to the mean outcome of a single trial and standard deviation proportional to $\frac{1}{\sqrt{n}}$.	If two events are disjoint, the probability that one or the other occurs is the sum of their individual probabilities.
The probability that an event does not occur is always one minus the probability that it does occur.	The branch of mathematics that counts arrangements of objects.
A curve that is always on or above the horizontal axis and has area exactly 1 underneath it. A density curve describes a continuous probability model.	A probability model that assigns probabilities to events as areas under a density curve.
Events that have no outcomes in common.	A probability model that assigns probabilities to each of a finite number of possible outcomes.

Chapter 8 Probability: The Mathematics of Chance **Event**	Chapter 8 Probability: The Mathematics of Chance **Law of large numbers**
Chapter 8 Probability: The Mathematics of Chance **Mean of a probability model**	Chapter 8 Probability: The Mathematics of Chance **Probability**
Chapter 8 Probability: The Mathematics of Chance **Probability histogram**	Chapter 8 Probability: The Mathematics of Chance **Probability model**
Chapter 8 Probability: The Mathematics of Chance **Random phenomenon**	Chapter 8 Probability: The Mathematics of Chance **Sample space**

As a random phenomenon is repeated many times, the mean $\overline{x}$ of the observed outcomes approaches the mean μ of the probability model.

Any collection of possible outcomes of a random phenomenon. An event is a subset of the sample space.

A number between 0 and 1 that gives the long-run proportion of repetitions of a random phenomenon on which an event will occur.

The average outcome of a random phenomenon with numerical values. When possible values $x_1, x_2, \ldots, x_k$ have probabilities $p_1, p_2, \ldots, p_k$, the mean is the average of the outcomes weighted by their probabilities, $\mu = x_1 p_1 + x_2 p_2 + \ldots + x_k p_k$.

A sample space S together with an assignment of probabilities to events. The two main types of probability models are discrete and continuous.

A histogram that displays a discrete probability model when the outcomes are numerical. The height of each bar is the probability of the outcome or group of outcomes at the base of the bar.

A list of all possible outcomes of a random phenomenon.

A phenomenon is random if it is uncertain what the next outcome will be, but each outcome nonetheless tends to occur in a fixed proportion of a very long sequence of repetitions. These long-run proportions are the probabilities of the outcomes.

Chapter 8
Probability: The Mathematics of Chance

Sampling distribution

Chapter 8
Probability: The Mathematics of Chance

Standard deviation of a probability model

A measure of the variability of a probability model. When the possible values $x_1, x_2, \ldots, x_k$ have probabilities $p_1, p_2, \ldots, p_k$, the variance is the average (weighted by probabilities) of the squared deviations from the mean, $\sigma^2 = (x_1 - \mu)^2 p_1 + (x_2 - \mu)^2 p_2 + \ldots + (x_k - \mu)^2 p_k.$ The standard deviation σ is the square root of the variance.

The distribution of values taken by a statistic when many random samples are drawn under the same circumstances. A sampling distribution consists of an assignment of probabilities to the possible values of a statistic.

Practice Quiz

1. We will roll a die and flip two coins. Then we will report the number on the die and whether the coins are heads, tails, or mixed. How many outcomes are in the sample space?

   a. 8

   b. 9

   c. 18

2. Suppose we toss three coins and report the number of heads that appear. What is the probability of exactly two heads appearing?

   a. $\dfrac{3}{8}$

   b. $\dfrac{1}{3}$

   c. $\dfrac{1}{2}$

3. A sample space contains three outcomes: A, B, C. Which of the following could be a legitimate assignment of probabilities to the outcomes?

   a. $P(A) = 0.4$ $P(B) = 0.6$ $P(C) = 0$

   b. $P(A) = 0.3$ $P(B) = 0.3$ $P(C) = 0.3$

   c. $P(A) = 0.6$ $P(B) = -0.2$ $P(C) = 0.6$

4. We roll two dice and report the sum of the numbers rolled. The outcomes in this space are

   a. all equally likely.

   b. not all equally likely.

5. A bicycle chain has a 4-digit code lock. How many possible codes are there if digits can be repeated?

   a. 40

   b. 5,000

   c. 10,000

6. A bicycle chain has a 4-digit code lock. How many possible codes are there if digits can't be repeated?

   a. 34

   b. 5,040

   c. 10,000

7. Each raffle ticket costs $2. Of 400 tickets sold, one will win $250, and two others will each win $25. What is your mean value for one play?

   a.   −$1.25

   b.   −$0.75

   c. $0.75

8. Suppose a random number generator produces numbers between 0 and 5. What is the probability of an outcome that is **not** between 1.1 and 3.2?

   a.   0.65625

   b.   0.42

   c.   0.58

9. Given the following probability model, find the mean and standard deviation.

Outcome	1	3	5
Probability	0.2	0.5	0.3

   a.   3.2; 1.4

   b.   3.2; 77.788

   c.   3; 2

10. The grades on a college-wide marketing exam were normally distributed with $\mu = 68.9$ and $\sigma = 7.1$. Given a SRS of 16 students who took the exam, what is the approximate probability that the mean score $\bar{x}$ of these 16 students is 72.45 or higher?

   a.   0.003

   b.   0.025

   c.   0.5

Word Search

Refer to pages 323 – 324 of your text to obtain the Review Vocabulary. There are 16 hidden vocabulary words/expressions in the word search below. *Continuous probability model* and *Standard deviation of a probability mode* do not appear in the word search due to expression length. The word *Probability* and *Event* appear separately from the other phrases that contain these words in the word search. It should be noted that spaces are removed.

```
A D R T I M R O E D N D E M R V E H M Z E A E G U
J O C M X T G G N J X C I E L A X F A R S A E R O
C O M B I N A T O R I C S A E E T L G B H T L P V
B S M N C I H T E S G O L N Z O A M X Y T O W F C
N C I D V P O I W R A N D O M P H E N O M E N O N
P O E D L E Y Z S N P P S F Z W E R E U O S T A D
R M Q H H A W A Z P R M A W R H O J P J O Y R O
F P F P X W J C M N R O H P N T R E R W R V A V G
A L R R S F S O P E O B J R P E E H O L E F N A N
Z E K O J Y T I L I B A B O R P E T E R C S I D G
E M R B I M N Z I E A B S B E R I T A I A L E D S
F E P B N R E S N E B I E A H A Y I A D P B E I J
P N I B G E V R G I I L M B E F D M O E S V D T Q
X T I I D S E S D J L I X I G I P I E E E L K I E
P R G L E G T K I T I T F L N D A L B N L E P O D
F U S I E H N M S R T Y E I Z T A L T I P N I N J
E L J T D E I I T Q Y H E T C O A A S W M E J R J
O E F Z I S O A R S M I F Y N C J R E M A I T U B
I D O N S I J X I I O S M M M T G T E A S Q A L O
L P V X U N S P B G D T T O E Q N N K A A F A E N
I O H S E E I M U T E O M D I O T E H G J L U R S
S B F T Z E D Q T Z L G F E V R U C Y T I S N E D
P E Y M A G E P I E H R X L T G E T S Q E I J M R
D E A G N L A W O F L A R G E N U M B E R S D T M
P M I E V E U E N E O M O H X N H R R R T H N E F
```

1. _____   9. _____

2. _____   10. _____

3. _____   11. _____

4. _____   12. _____

5. _____   13. _____

6. _____   14. _____

7. _____   15. _____

8. _____   16. _____

Chapter 9
Social Choice: The Impossible Dream

Chapter Objectives

Check off these skills when you feel that you have mastered them.

☐ Analyze and interpret preference list ballots.

☐ Explain three desired properties of Majority Rule.

☐ Explain May's theorem.

☐ Explain the difference between majority rule and the plurality method.

☐ Discuss why the majority method may not be appropriate for an election in which there are more than two candidates.

☐ Apply the plurality voting method to determine the winner in an election whose preference list ballots are given.

☐ Explain the Condorcet winner criterion (CWC).

☐ Rearrange preference list ballots to accommodate the elimination of one or more candidates.

☐ Structure two alternative contests from a preference schedule by rearranging preference list ballots; then determine whether a Condorcet winner exists.

☐ Apply the Borda count method to determine the winner from preference list ballots.

☐ Explain independence of irrelevant alternatives (IIA).

☐ Apply the sequential pairwise voting method to determine the winner from preference list ballots.

☐ Explain Pareto condition.

☐ Apply the Hare system to determine the winner from preference list ballots

☐ Explain monotonicity.

☐ Apply the plurality runoff method to determine the winner from preference list ballots.

☐ Explain Arrow's impossibility theorem.

☐ Recognize the application of the law of transitivity in the interpretation of individual preference schedules and its possible nonvalidity in group preferences (the Condorcet paradox).

☐ Apply the process of approval voting and discuss its consideration in political races.

Guided Reading

Introduction

Voting occurs in many situations, such as in elections of public officials, officers of a club, or among a group of friends who have to decide in which restaurant to eat. While elections involving just two choices are quite simple, the opposite is true for elections with three or more alternatives, in which many complications and paradoxes arise. Social choice theory was developed to analyze the various types of voting methods, to discover the potential pitfalls in each, and to attempt to find improved systems of voting.

In this chapter, you will be interpreting and altering preference list ballots. A **preference list ballot** consists of a rank ordering of candidates. Throughout the chapter these vertical lists will have the most preferential candidate on top and the least preferential on the bottom. An example of a voter's ballot for four candidates (say A, B, C, and D) could be as follows.

Rank	
First	A
Second	D
Third	B
Fourth	C

For a particular vote, we can summarize the preference list ballots in a single table as follows.

Rank	Number of voters (15)			
	5	2	7	1
First	A	C	B	D
Second	D	B	C	A
Third	B	A	D	B
Fourth	C	D	A	C

For this example, there are four candidates (namely A, B, C, and D). There are a total of 15 voters. The number above the four different preference list ballots represents how many voters identified that particular column as the ordering of the candidates as their preference.

Throughout this chapter it will be assumed that there are an **odd number of voters** for any given discussion.

Section 9.1   Majority Rule and Condorcet's Method

☞ **Key idea**

In a **dictatorship** all ballots except that of the dictator are ignored.

☞ **Key idea**

In **imposed rule** candidate X wins regardless of who vote for whom.

☞ **Key idea**

In **minority rule**, the candidate with the fewest votes wins.

☞ **Key idea**

When there are only two candidates or alternatives, May's theorem states that **majority rule** is the only voting method that satisfies three desirable properties, given an odd number of voters and no ties.

✎ Question 1

What are three properties satisfied by majority rule?

Answer

The three properties are:

1. All voters are treated equally.
2. Both candidates are treated equally.
3. If a single voter who voted for the loser, B, changes his mind and votes for the winner, A, then A is still the winner. This is what is called **monotone**.

⚷ Key idea

Condorcet's method declares a candidate is a winner if he or she can defeat every other candidate in a one-on-one competition using majority rule.

✑ Example A

Determine if there is a winner using Condorcet's method. If so, who is it?

Rank	Number of voters (11)			
	2	5	3	1
First	A	B	B	C
Second	B	A	C	A
Third	C	C	A	B

Solution

You must determine the outcome of three one-on-one competitions. The candidates not considered in each one-on-one competition can be ignored.

A vs B

Rank	Number of voters (11)			
	2	5	3	1
First	A	B	B	A
Second	B	A	A	B

A: $2 + 1 = 3$; B: $5 + 3 = 8$

A vs C

Rank	Number of voters (11)			
	2	5	3	1
First	A	A	C	C
Second	C	C	A	A

A: $2 + 5 = 7$; C: $3 + 1 = 4$

B vs C

Rank	Number of voters (11)			
	2	5	3	1
First	B	B	B	C
Second	C	C	C	B

B: $2 + 5 + 3 = 10$; C: 1

Since B can defeat both A and C in a one-on-one competition, B is the winner by the Condorcet's method.

✎ Question 2

Determine if there is a winner using Condorcet's method. If so, who is it?

	Number of voters (15)			
Rank	5	2	7	1
First	A	C	B	B
Second	B	B	C	A
Third	C	A	A	C

Answer

B is the winner.

✎ Question 3

In the following table, is there a Condorcet winner? If so, who is it?

	Number of voters (23)			
Rank	3	8	7	5
First	A	B	C	D
Second	B	C	B	C
Third	C	A	A	B
Fourth	D	D	D	A

Answer

C is the winner.

⚷ Key idea

Condorcet's voting paradox *can* occur with three or more candidates in an election where Condorcet's method yields no winners. For example, in a three-candidate race, two-thirds of voters could favor *A* over *B*, two-thirds of voters could favor *B* over *C*, and two-thirds of voters could favor *C* over *A*. This is the example given in the text. With three or more candidates, there are elections in which Condorcet's method yields no winners.

✐ Example B

Does Condorcet's voting paradox occur in the following table?

	Number of voters (9)		
Rank	3	4	2
First	A	B	C
Second	C	A	B
Third	B	C	A

Solution

Yes, the Condorcet's voting paradox occurs. Voters prefer *A* over *C* (7 to 2). Voters prefer *C* over *B* (5 to 4). However, voters prefer *B* over *A* (6 to 3).

ᕐ Example C

Does Condorcet's voting paradox occur in the following table?

	Number of voters (13)			
Rank	5	2	4	2
First	A	B	C	D
Second	B	C	D	B
Third	D	A	A	A
Fourth	C	D	B	C

Solution

The answer is no. Voters prefer A over B (9 to 4). Voters prefer A over C (7 to 6). Also, voters prefer A over D (7 to 6). A is the Condorcet winner. Therefore, there is no Condorcet paradox.

✎ Question 4

Does Condorcet's voting paradox occur in the following table?

	Number of voters (23)			
Rank	9	4	2	8
First	A	B	B	C
Second	B	C	A	A
Third	C	A	C	B

Answer

The answer is yes.

Section 9.2   Other Voting Systems for Three of More Candidates

⌗ Key idea

In **plurality voting**, the candidate with the most first-place votes on the preference list ballots is the winner. We do not take into account the voters' preferences for the second, third, etc., places.

ᕐ Example D

In the following table, who is the winner by plurality voting?

	Number of voters (23)			
Rank	7	8	6	2
First	A	C	B	A
Second	B	B	C	C
Third	C	A	A	B

Solution

Since we only need to consider first-place votes, we have the following.

	Number of voters (23)			
Rank	7	8	6	2
First	A	C	B	A

Thus, A has 7 + 2 = 9 first-place votes, B has 6 and C has 8. Thus, A is the winner.

⌐ Key idea

A voting system that satisfies the **Condorcet Winner Criterion (CWC)** either has no Condorcet winner or the voting produces exactly the same winner as does Condorcet's method.

✎ Question 5

In the following table, who is the winner by plurality voting? Is there a winner by Condorcet's method? Is there a violation of CWC?

	Number of voters (27)				
Rank	11	2	8	2	4
First	A	A	B	C	C
Second	B	C	C	A	B
Third	C	B	A	B	A

Answer

The winner is *A*; There is no winner.; There is no violation.

✎ Question 6

In plurality voting, must there always be a Condorcet winner?

Answer

The answer is no.

⌐ Key idea

In the **Borda count** method, points are assigned to each position in the set of preference lists. For example, in a 3-person election, first-place votes may be awarded 2 points each, second-place votes receive 1 point each, and third-place votes are given 0 points each. (Other distributions of points may be used to create similar rank methods.)

ᴑ Example E

In the following table, who is the winner by Borda count?

	Number of voters (27)				
Rank	11	2	8	2	4
First	A	A	B	C	C
Second	B	C	C	A	B
Third	C	B	A	B	A

Solution

Preference	1st place votes × 2	2nd place votes × 1	3rd place votes × 0	Borda score
A	13 × 2	2 × 1	12 × 0	28
B	8 × 2	15 × 1	4 × 0	31
C	6 × 2	10 × 1	11 × 0	22

The winner is *B*.

⌐ Key idea

Another method of determining Borda scores outlined by the text is to individually replace the candidates below the one you are determining the score for by a box. You then **count up the boxes**, being careful to take note of the number of voters in each column.

⌐ Example F

Use the preference list ballots from the previous example to show the Borda scores are 28 for A, 31 for B, and 22 for C using the box-replacement method.

Solution

For A:

Rank	Number of voters (27)				
	11	2	8	2	4
First	A	A	B	C	C
Second	☐	☐	C	A	B
Third	☐	☐	A	☐	A

The Borda score for A is $(11 \times 2) + (2 \times 2) + (2 \times 1) = 22 + 4 + 2 = 28$.

For B:

Rank	Number of voters (27)				
	11	2	8	2	4
First	A	A	B	C	C
Second	B	C	☐	A	B
Third	☐	B	☐	B	☐

The Borda score for B is $(11 \times 1) + (8 \times 2) + (4 \times 1) = 11 + 16 + 4 = 31$.

For C:

Rank	Number of voters (27)				
	11	2	8	2	4
First	A	A	B	C	C
Second	B	C	C	☐	☐
Third	C	☐	☐	☐	☐

The Borda score for C is $(2 \times 1) + (8 \times 1) + (2 \times 2) + (4 \times 2) = 2 + 8 + 4 + 8 = 22$.

⌐ Key idea

In the Borda count method, you can perform a check to make sure your calculations are correct. The number of points to be distributed per preference list ballot times the number of ballots (voters) is equal to the sum of the Borda scores. For example in the previous table, 3 points $(2 + 1 + 0 = 3)$ were to be distributed among 3 candidates. There were 27 voters. The sum of the Borda scores was $28 + 31 + 22 = 81$. This is the same as the product of 3 and 27. If there are four candidates, you will be distributing 6 points $(3 + 2 + 1 + 0 = 6)$ using the point distribution of the text.

✎ Question 7

In the following table, who is the winner by Borda count? What is the sum of all the Borda scores?

	Number of voters (37)					
Rank	8	8	2	12	5	2
First	A	A	B	C	C	D
Second	B	D	C	B	D	A
Third	C	B	D	A	A	B
Fourth	D	C	A	D	B	C

Answer

The winner is A.; 222

⌐ Key idea

A voting system satisfies **independence of irrelevant alternatives (IIA)** if it is impossible for a candidate B to move from nonwinner status to winner status unless at least one voter reverses the order in which he or she had B and the winning candidate ranked. The Borda count fails to satisfy IIA, as shown in the text.

⌐ Key idea

An agenda is the listing (in some order) of the candidates. **Sequential pairwise** voting pits the first candidate against the second in a one-on-one contest. The winner goes on to confront the third candidate on the agenda, while the loser is eliminated. The candidate remaining at the end is the winner. The choice of the agenda can affect the result.

ᏻ Example G

Who is the winner with sequential pairwise voting with the agenda B, C, A?

	Number of voters (17)			
Rank	1	5	4	7
First	A	A	B	C
Second	B	C	A	A
Third	C	B	C	B

Solution

In sequential pairwise voting with the agenda B, C, A, we first pit B against C. There are 5 voters who prefer B to C and 12 prefer C to B. Thus, C wins by a score of 12 to 5. B is therefore eliminated, and C moves on to confront A. There are 7 voters who prefer C to A and 10 prefer A to C. Thus, A wins by a score of 10 to 7. Thus, A is the winner by sequential pairwise voting with the agenda B, C, A.

✏ Question 8

Who is the winner with sequential pairwise voting with the agenda A, B, C, D? with agenda B, D, C, A?

Rank	\multicolumn{5}{c}{Number of voters (19)}				
	5	3	2	1	8
First	A	A	B	C	D
Second	B	D	C	B	B
Third	D	B	A	A	A
Fourth	C	C	D	D	C

Answer

D is the winner.; A is the winner.

🗝 Key idea

Sequential pairwise voting fails to satisfy the **Pareto condition**, which states that if everyone prefers one candidate, say A, to another, say B, then B cannot be the winner.

🗝 Key idea

In the **Hare system**, the winner is determined by repeatedly deleting candidates that are the least preferred, in the sense of being at the top of the fewest preference lists.

✍ Example H

Who is the winner when the Hare system is applied?

Rank	\multicolumn{4}{c}{Number of voters (19)}			
	3	5	4	7
First	A	A	B	C
Second	B	C	C	A
Third	C	B	A	B

Solution

Since B has the least number of first-place votes, B is eliminated.

Rank	\multicolumn{4}{c}{Number of voters (19)}			
	3	5	4	7
First	A	A		C
Second		C	C	A
Third	C		A	

Candidates A and C move up as indicated to form a new table.

Rank	\multicolumn{4}{c}{Number of voters (19)}			
	3	5	4	7
First	A	A	C	C
Second	C	C	A	A

A now has $3 + 5 = 8$ first-place votes. C now has $4 + 7 = 11$ first-place votes. Thus, C is the winner by the Hare system.

✎ Question 9

Who is the winner when the Hare system is applied?

Rank	\multicolumn{5}{Number of voters (17)}				
	5	2	6	3	1
First	A	B	C	D	B
Second	B	A	A	C	D
Third	C	C	B	B	C
Fourth	D	D	D	A	A

Answer

C is the winner.

✎ Question 10

Who is the winner when the Hare system is applied?

Rank	Number of voters (21)				
	1	8	7	3	2
First	A	B	C	D	E
Second	B	E	A	A	A
Third	C	C	B	C	C
Fourth	D	D	D	B	B
Fifth	E	A	E	E	D

Answer

C is the winner.

⚷ Key idea

The Hare system does not satisfy monotonicity.

⚷ Key idea

Plurality runoff is the voting system in which there is a runoff between the two candidates receiving the most first-place votes. In the case of ties between first or second, three candidates participate in the runoff.

🖎 Example I

Who is the winner when the plurality runoff method is applied?

	Number of voters (21)				
Rank	7	3	2	1	8
First	*A*	*B*	*C*	*D*	*D*
Second	*C*	*A*	*A*	*C*	*B*
Third	*B*	*C*	*B*	*B*	*C*
Fourth	*D*	*D*	*D*	*A*	*A*

Solution

Since *D* and *A* are in first and second-place, respectively, the runoff is between these two candidates.

	Number of voters (21)				
Rank	7	3	2	1	8
First	*A*	*A*	*A*	*D*	*D*
Second	*D*	*D*	*D*	*A*	*A*

A now has $7 + 3 + 2 = 12$ first-place votes, and *D* has $1 + 8 = 9$. Thus, *A* is the winner.

✏ Question 11

Who is the winner when the plurality runoff method is applied?

	Number of voters (29)				
Rank	9	6	7	2	5
First	*A*	*B*	*C*	*D*	*D*
Second	*C*	*D*	*B*	*C*	*B*
Third	*B*	*A*	*A*	*A*	*C*
Fourth	*D*	*C*	*D*	*B*	*A*

Answer

D is the winner.

☞ Key idea

The Plurality runoff method does not satisfy monotonicity.

Section 9.3   Insurmountable Difficulties: Arrow's Impossibility Theorem

☞ Key idea

Arrow's impossibility theorem states that with three or more candidates and any number of voters, there does not exist (and will never exist) a voting system that produces a winner satisfying Pareto and independence of irrelevant alternatives (IIA), and is not a dictatorship.

☞ Key idea

A weak version of Arrow's impossibility theorem states that with three or more candidates and an odd number of voters, there does not exist (and will never exist) a voting system that satisfies both the Condorcet winner criterion (CWC) and independence of irrelevant alternatives (IIA), and that always produces at least one winner every election.

Section 9.4   A Better Approach? Approval Voting

⌐ Key idea

In **approval voting**, each voter may vote for as many candidates as he or she chooses. The candidate with the highest number of approval votes wins the election.

⌐ Example J

There are 7 voters in a committee. Who is the winner in the following table, where X indicates that the voter approves of that particular candidate? How would they be ranked?

| | Voters | | | | | | |
Candidates	1	2	3	4	5	6	7
A			X	X		X	
B		X	X		X	X	
C				X			X
D		X	X	X	X		X
E			X		X		X

Solution

A has 3 approval votes, B has 4, C has 2, D has 5, E has 3. Since D has the most approval votes, D is the winner. Ranking the candidates we have, D (5), B (4), A and E (3), and C (2).

⌐ Example K

There are 25 voters in a committee. Who is the winner in the following table, where X indicates that the voter approves of that particular candidate? How would they be ranked?

| | Number of voters (25) | | | | | | |
Nominee	5	8	6	2	1	1	2
A	X			X	X		X
B		X		X		X	X
C	X		X			X	X

Solution

A has $5+2+1+2=10$ approval votes.   B has $8+2+1+2=13$ approval votes.   C has $5+6+1+2=14$ approval votes. Since C has the most approval votes, C is the winner. Ranking the candidates we have, C (14), B (13), and A (10).

✎ Question 12

There are 71 voters in a committee. Who is the winner in the following table, where X indicates that the voter approves of that particular candidate? How would they be ranked?

| | Number of voters (71) | | | | | | | |
Nominee	8	9	2	12	10	14	12	4
A	X			X	X			X
B		X		X		X		X
C						X		X
D		X		X		X		X
E	X	X	X		X	X		X

Answer

E is the winner.; Ranking is E, B and D (tie), A, and C.

Homework Help

Exercises 1 – 3
Carefully read Section 9.1 before responding to these exercises. Consider all three desirable properties for each exercise.

Exercise 4
Consider 4 voters (an even number of voters) and 2 candidates. Name one of the voters to distinguish him or her from the others.

Exercise 5
Consider what cannot occur when you have an odd number of voters.

Exercise 6
Consider preference choices like softdrinks, courses, makes of cars, etc.

Exercise 7
(a) Check the one-on-one scores of D versus H, D versus J, and H versus J.
(b) The plurality winner would be the candidate with the highest percentage of votes in this exercise.

Exercises 8, 9, and 12
(a) The plurality winner would be the candidate with the highest number of votes in these exercises.
(b) Use the following table for these exercises unless the "counting box" method is preferred. Be sure to check the total of the Borda scores after you have completed the table.

Preference	1st place votes × 3	2nd place votes × 2	3rd place votes × 1	4th place votes × 0	Borda score
A	×3	×2	×1	×0	
B	×3	×2	×1	×0	
C	×3	×2	×1	×0	
D	×3	×2	×1	×0	

(c) Eliminate the candidate(s) with the least number of first-place votes. Repeat if necessary until there are two candidates. The one with the majority of votes wins.
(d) The exercises have different agendas. Be sure to pay attention to the agenda in determining the three one-on-one competitions.

Exercises 10, 11, and 13
(a) The plurality winner would be the candidate with the highest number of votes in these exercises.
(b) Use the following table for these exercises unless the "counting box" method is preferred. Be sure to check the total of the Borda scores after you have completed the table.

Preference	1st place votes × 4	2nd place votes × 3	3rd place votes × 2	4th place votes × 1	5th place votes × 0	Borda score
A	×4	×3	×2	×1	×0	
B	×4	×3	×2	×1	×0	
C	×4	×3	×2	×1	×0	
D	×4	×3	×2	×1	×0	
E	×4	×3	×2	×1	×0	

(c) Eliminate the candidate(s) with the least number of first-place votes. Repeat if necessary until there are two candidates. The one with the majority of votes wins.
(d) The exercises have different agendas. Be sure to pay attention to the agenda in determining four one-on-one competitions.

Exercise 14

(a) The plurality winner would be the candidate with the highest number of votes in these exercises.

(b) Use the following table for these exercises unless the "counting box" method is preferred. Be sure to check the total of the Borda scores after you have completed the table.

Preference	1st place votes × 4	2nd place votes × 3	3rd place votes × 2	4th place votes × 1	5th place votes × 0	Borda score
A	×4	×3	×2	×1	×0	
B	×4	×3	×2	×1	×0	
C	×4	×3	×2	×1	×0	
D	×4	×3	×2	×1	×0	
E	×4	×3	×2	×1	×0	

(c) This exercise has A, B, C, D, E as its agenda. Be sure to pay attention to the agenda in determining four one-on-one competitions.

(d) Eliminate the candidate(s) with the least number of first-place votes. Repeat if necessary until there are two candidates. The one with the majority of votes wins.

Exercise 15

(a) You should find the following tables helpful in the elimination process. Make sure you use the Coombs procedure.

Rank	Number of voters (7)						
	1	1	1	1	1	1	1
First							
Second							
Third							
Fourth							

Rank	Number of voters (7)						
	1	1	1	1	1	1	1
First							
Second							
Third							

Rank	Number of voters (7)						
	1	1	1	1	1	1	1
First							
Second							

(b) Consider one possible scenario is having candidates A, B, and C. Fill in the preference list ballots and determine the winner using both the Coombs procedure and the Hare method. Make sure your example fits the criteria in the exercise.

Rank	Number of voters ( )
First	
Second	
Third	

Exercise 16

(a) Read carefully the voting procedure known as the Condorcet's method (Section 9.1) and the Pareto condition (Section 9.2).

(b) Read carefully the voting procedure known as the Condorcet's method (Section 9.1) and the monotonicity condition (Section 9.2).

Exercise 17
(a) Read carefully the voting procedure known as plurality voting (Section 9.2) and the Pareto condition (Section 9.2).
(b) Read carefully the voting procedure known as plurality voting (Section 9.2) and the monotonicity condition (Section 9.2).

Exercise 18
(a) Read carefully the voting procedure known as the Borda method (Section 9.2) and the Pareto condition (Section 9.2).
(b) Read carefully the voting procedure known as the Borda method (Section 9.2) and the monotonicity condition (Section 9.2).

Exercise 19
(a) Read carefully the voting procedure known as the sequential pairwise voting method (Section 9.2) and the Condorcet winner criterion condition (Section 9.2).
(b) Read carefully the voting procedure known as the sequential pairwise voting method (Section 9.2) and the monotonicity condition (Section 9.2).

Exercise 20
Read carefully the voting procedure known as the Hare system (Section 9.2) and the Pareto condition (Section 9.2).

Exercise 21
Read carefully the voting procedure known as the plurality runoff method (Section 9.2) and the Pareto condition (Section 9.2).

Exercise 22
Determine the winner using the plurality runoff method. Determine if there is a Condorcet winner. Discuss the outcome.

Exercise 23
Adjust the table in such a way that the change is favorable to the winning candidate. Determine the winner using the plurality runoff method on the adjusted table. Discuss the outcome.

Exercise 24
(a) Determine the winner using plurality voting in the first election, then in the second. Determine the changes that were made from the first to the second election and the effect on the second election. Be sure to have reviewed the independence of irrelevant alternatives (Section 9.2) before making your observations.
(b) Determine the winner using the Hare system in the first election, then in the second. Determine the changes that were made from the first to the second election and the effect on the second election. Be sure to have reviewed the independence of irrelevant alternatives (Section 9.2) before making your observations.

Exercise 25
Try to keep your example simple, say with three candidates and a relatively small number of voters.

	Number of voters ( )
Rank	
First	
Second	
Third	

Determine the winner using the Borda method.

Preference	1st place votes × 2	2nd place votes × 1	3rd place votes × 0	Borda score
A	× 2	× 1	× 0	
B	× 2	× 1	× 0	
C	× 2	× 1	× 0	

Now use Condorcet's method with your preference list ballots. You have shown the Borda count does not satisfy the Condorcet winner criterion if you obtain two different winners.

Exercise 26
Determine the winner using the Hare system. You should find the following table helpful in the elimination process.

	Number of voters (17)			
Rank	7	5	4	1
First				
Second				
Third				

Alter the original table to an outcome favorable to the winner.

	Number of voters (17)			
Rank	7	5	4	1
First				
Second				
Third				
Fourth				

Using the Hare system again on the altered table.

	Number of voters (17)			
Rank	7	5	4	1
First				
Second				
Third				

	Number of voters (17)			
Rank	7	5	4	1
First				
Second				

You will have demonstrated nonmonotonicity if the outcome of the altered table is not the same as the original winner, even though the alteration was favorable to the original winner.

Exercise 27

(a) Determine the winner using the Hare system. You should find the following tables helpful in the elimination process.

Rank	Number of voters (21)			
	7	6	5	3
First				
Second				
Third				

Rank	Number of voters (21)			
	7	6	5	3
First				
Second				

(b) Alter the table as described in the exercise.

Rank	Number of voters (21)			
	7	6	5	3
First				
Second				
Third				
Fourth				

Determine the winner again using the Hare system. You should find the following tables helpful in the elimination process.

Rank	Number of voters (21)			
	7	6	5	3
First				
Second				
Third				

Rank	Number of voters (21)			
	7	6	5	3
First				
Second				

Exercise 28

(a) Read carefully the voting procedure known as sequential pairwise voting (Section 9.2) and consider what cannot occur when you have an odd number of voters.

(b) Read carefully the voting procedure known as the Hare system (Section 9.2) and consider what cannot occur when you have an odd number of voters.

Exercise 29

Consider constructing preference list ballots that rank a candidate (say C) last on a majority of ballots (over 50% of the votes are for last-place). Preference list ballots that have three candidates (say A, B, and C) and a relatively small number of voters (say 11) will suffice. Determine what occurs using the plurality runoff method. Discuss in general what would occur if a candidate was to be ranked last on a majority of votes. Note that it is assumed that there are no ties for first or second place in this exercise.

Exercise 30
Consider constructing preference list ballots that rank a candidate with three candidates (say A, B, and C) and a relatively small number of voters. Force one candidate to have the majority of the last-place votes, but is the winner by the plurality voting (i.e. has the highest number of first-place votes).

	Number of voters ( )		
Rank	3	2	2
First			
Second			
Third			

Exercise 31
Follow the argument in the text in Section 9.3. Consider starting with the following preference list ballots for the case involving B.

Rank	Number of voters (3)		
First	A	B	A
Second	B	C	C
Third	C	A	B

Consider starting with the following preference list ballots for the case involving C.

Rank	Number of voters (3)		
First	B	B	C
Second	A	C	A
Third	C	A	B

Exercise 32

In the table, the numbers $1 - 10$ identify the ten board members. For each candidate ($A - H$) count horizontally the number of X marks, then answer the questions, Parts a – d.

Exercise 33
Similar to the preference list ballots, the numbers in the second row correspond to how many voters chose to approve a combination of nominees. If a nominee has an X under a number of voters, you add the number of voters indicated to find the total number of approval votes out of the possible 45. For example, A has $7+9+6+1=23$ approval votes. Find the number of approval votes for the other two nominees in order to answer Parts a and b. In Part c, consider what effect the last two columns have on the outcome.

Do You Know the Terms?

Cut out the following 19 flashcards to test yourself on Review Vocabulary. You can also find these flashcards at http://www.whfreeman.com/fapp7e.

Chapter 9 Social Choice: The Impossible Dream **Agenda**	Chapter 9 Social Choice: The Impossible Dream **Approval voting**
Chapter 9 Social Choice: The Impossible Dream **Arrow's impossibility theorem**	Chapter 9 Social Choice: The Impossible Dream **Borda count**
Chapter 9 Social Choice: The Impossible Dream **Condorcet's Method**	Chapter 9 Social Choice: The Impossible Dream **Condorcet winner**
Chapter 9 Social Choice: The Impossible Dream **Condorcet winner criterion (CWC)**	Chapter 9 Social Choice: The Impossible Dream **Condorcet's voting paradox**
Chapter 9 Social Choice: The Impossible Dream **Hare system**	Chapter 9 Social Choice: The Impossible Dream **Independence of irrelevant alternatives (IIA)**

A method of electing one or more candidates from a field of several in which each voter submits a ballot that indicates which candidates he or she approves of. Winning is determined by the total number of approvals a candidate obtains.

An ordering of the candidates for consideration. Often used in sequential pairwise voting.

A voting system for elections with several candidates in which points are assigned to voters' preferences and these points are summed for each candidate to determine a winner.

Kenneth J. Arrow's discovery that any voting system can give undesirable outcomes.

A Condorcet winner in an election is a candidate who, based on the ballots, would have defeated every other candidate in a one-on-one contest.

A voting system for elections with several candidates in which a candidate is a winner precisely when he or she would, on the basis of the ballots, defeat every other candidate on a one-on-one contest.

There are elections in which Condorcet's method yields no winner.

A voting system satisfies the Condorcet winner criterion if, for every election in which there is a Condorcet winner, it wins the election when that voting system is used.

A voting system satisfies independence of irrelevant alternatives if the only way a candidate (call him A) can go from losing one election to being among the winners of a new election (with the same set of candidates and voters) is for at least one voter to reverse his or her ranking

A voting system for elections with several candidates in which candidates are successively eliminated in an order based on the number of first-place votes.

Chapter 9
Social Choice: The Impossible Dream

Manipulability

Chapter 9
Social Choice: The Impossible Dream

Majority rule

Chapter 9
Social Choice: The Impossible Dream

May's theorem

Chapter 9
Social Choice: The Impossible Dream

Monotonicity

Chapter 9
Social Choice: The Impossible Dream

Pareto condition

Chapter 9
Social Choice: The Impossible Dream

Plurality runoff

Chapter 9
Social Choice: The Impossible Dream

Plurality voting

Chapter 9
Social Choice: The Impossible Dream

Preference list ballot

Chapter 9
Social Choice: The Impossible Dream

Sequential pairwise voting

A voting system for elections with two candidates (and an odd number of voters) in which the candidate preferred by more than half the voters is the winner.	A voting system is subject to manipulability id there are elections in which it is to a voter's advantage to submit a ballot that misrepresents his or her true preferences.
A voting system satisfies monotonicity provided that ballot changes favorable to one candidate (and not favorable to any other candidate) can never hurt that candidate.	Kenneth May's discovery that, for two alternatives and an odd number of voters, majority rule is the only voting system satisfying three natural properties.
A voting system for elections with several candidates in which, assuming there are no ties, there is a runoff between the two candidates receiving the most first-place votes.	A voting system satisfies the Pareto condition provided that every voter's ranking of one candidate higher than another precludes the possibility of this latter candidate winning.
A ballot that ranks the candidates from most preferred to least preferred with no ties	A voting system for elections with several candidates in which the candidate with the most first-place votes wins.
	A voting system for elections with several candidates in which one starts with an agenda and pits the candidates against each other in one-on-one contests (based on preference list ballots), with losers being eliminated as one moves along the agenda.

Practice Quiz

1. Majority rule is an effective way to make a choice between

   a. two alternatives.

   b. a small number of alternatives.

   c. any number of alternatives.

2. The first-place votes for each of four candidates are counted, and the candidate with the most votes wins. This voting system is an example of

   a. majority rule.

   b. approval voting.

   c. plurality voting.

3. Each voter ranks the four candidates. The candidate who is ranked above any of the other candidates by a majority of the voters is declared to be the winner. This is an example of

   a. Condorcet winner criterion.

   b. Borda count.

   c. Hare system.

4. 11 committee members need to elect a chair from the candidates *A*, *B*, *C*, and *D*. The preferences of the committee members are given below. Which candidate will be selected if they use majority rule?

	Number of voter members (11)		
Rank	**6**	**2**	**3**
First	*A*	*B*	*C*
Second	*B*	*C*	*D*
Third	*C*	*D*	*B*
Fourth	*D*	*A*	*A*

   a. *A*

   b. *B*

   c. *C*

5. 11 committee members need to elect a chair from the candidates *A*, *B*, *C*, and *D*. The preferences of the committee members are given below. Which candidate will be selected if they use a Borda count?

	Number of voter members (11)		
Rank	**6**	**2**	**3**
First	*A*	*B*	*C*
Second	*B*	*C*	*D*
Third	*C*	*D*	*B*
Fourth	*D*	*A*	*A*

   a. *A*

   b. *B*

   c. *C*

6. 11 committee members need to elect a chair from the candidates A, B, C, and D. The preferences of the committee members are given below. Which candidate will be selected if they use the plurality runoff?

	Number of voter members (11)		
Rank	**6**	**2**	**3**
First	A	B	C
Second	B	C	D
Third	C	D	B
Fourth	D	A	A

    a. A

    b. B

    c. C

7. 37 members must elect a club president. Preferences among candidates A, B, C, and D are given below. Which candidate wins under the Hare system?

	Number of voter members (37)				
Rank	**14**	**10**	**8**	**4**	**1**
First	A	C	D	B	A
Second	B	B	C	D	D
Third	C	D	B	C	B
Fourth	D	A	A	A	C

    a. A

    b. B

    c. D

8. 37 members must elect a president of their club. Preferences among candidates A, B, and C are given below. Which candidate is the Condorcet winner?

	Number of voter members (37)		
Rank	**14**	**11**	**12**
First	A	B	C
Second	B	C	B
Third	C	A	A

    a. A

    b. B

    c. C

9. 25 partygoers have enough money together for a one-topping super-size party pizza. They each mark what toppings they find acceptable, as shown below. Which topping will be selected using approval voting?

Toppings	Voters				
	8	6	4	4	3
pepperoni		X	X	X	
mushrooms				X	X
anchovies	X		X		X

a. pepperoni

b. mushrooms

c. anchovies

10. Which voting method satisfies both the Condorcet winner criterion and the independence of irrelevant alternatives?

I. Condorcet method   II. Plurality

a. Only I

b. Only II

c. Neither I nor II

Word Search

Refer to pages 361 – 362 of your text to obtain the Review Vocabulary. There are 17 hidden vocabulary words/expressions in the word search below. *Condorcet's method* and *Manipulability* do not appear. It should be noted that spaces are removed as well as apostrophes. Also, the abbreviations for Condorcet winner criterion and Independence of Irrelevant alternatives were used in the word search.

```
P L U R A L I T Y R U N O F F A I K X A S T F S S N
P L B E K P F I E G U C R P S E Y N I G X I U C Z C
R R I N M B L S G J M J J O G D C S A E H I D S Q X
A M B N I C S Y S B P H H N O P C M R N X A S J N K
P L A I V Z E P V T F L H X M A H E T D E M I E K A
S N A W Y K A F E X I S S D M N H E O A A N N N T M
G N I T O V L A V O R P P A T N U O C A D R O B A S
O B Q E C N G S N E E E J K H I G S S R M T F P G Y
A N Z C S W R M M O N O T O N I C I T Y F J O T A T
I Q F R E W C E P I R G N I T O V Y T I L A R U L P
M E R O E H T Y T I L I B I S S O P M I S W O R R A
C O N D O R C E T S V O T I N G P A R A D O X Y X R
P L A N R M A Y S T H E O R E M O C I O G E G W N E
T E P O B L R S E T V V S V N D S T L S G I O P R T
Z E G C R U M F B R S B T D R X R G H M M L D H D O
R S H F L A V I T R H D L P M J T M T Z N U E A F C
A L I E D A R C F E E P S T A A B P O S E S U R Q O
T O L L A B T S I L E C N E R E F E R P E K C E N N
G N I T O V E S I W R I A P L A I T N E U Q E S R D
J C Q C X T L N C A Y P O Y C E X F S N V E N Y L I
R X S N S T O K T N I H I O A C N M S N U A I S A T
F M M R C I N E T N P Z E N E C H W F S V P E T F I
F R R A P E X E R B F P O G T D U I I Q H A E E T O
N B S N A O H M E I N T E E O X L F U Q C E R M D N
C K U C N Z P T E P N E N E W H X N Y T P T N X A N
X W I Z N E H L Q A L H F D O O A T S J O Y J A J M
```

1. _____   10. _____

2. _____   11. _____

3. _____   12. _____

4. _____   13. _____

5. _____   14. _____

6. _____   15. _____

7. _____   16. _____

8. _____   17. _____

9. _____

Chapter 10
The Manipulability of Voting Systems

Chapter Objectives

Check off these skills when you feel that you have mastered them.

☐ Explain what is meant by voting manipulation.

☐ Determine if a voter, by a unilateral change, has manipulated the outcome of an election.

☐ Determine a unilateral change by a voter that causes manipulation of an election using the Borda count voting method.

☐ Explain the three conditions to determine if a voting system is manipulable.

☐ Discuss why the majority method may not be appropriate for an election in which there are more than two candidates.

☐ Explain four desirable properties of Condorcet's method.

☐ Explain why Condorcet's method is non-manipulable by a unilateral change in vote.

☐ Recognize when the Borda count method can be manipulated and when it can't.

☐ Determine a unilateral change by a voter that causes a no-winner manipulation of an election in Condorcet's method.

☐ Determine a unilateral change by a voter that causes manipulation of an election in the plurality runoff method.

☐ Determine a unilateral change by a voter that causes manipulation of an election in the Borda count voting method.

☐ Determine a unilateral change by a voter that causes manipulation of an election in the Hare method.

☐ Determine a group change by a block of voters that causes manipulation of an election in the plurality method.

☐ Determine an agenda change by a voter that causes manipulation of an election in the sequential pairwise voting method, with agenda.

☐ Explain the Gibbard-Satterthwaite theorem (GS theorem) and its weak version.

☐ Explain the chair's paradox and what is meant by *weakly dominates* as it relates to a voting strategy.

Guided Reading

Introduction

The expression, *Honesty is the best policy*, may not be applicable when it comes to voting. Voting in a strategic manner is called **manipulation**. This occurs when a voter casts a ballot, which does not represent his or her actual preference. These types of ballots are referred to as **insincere** or **disingenuous ballots**. In this chapter, you will be looking at the manipulability of different voting methods.

⚿ Key idea

In manipulating an outcome, a voter casts a vote that is not consistent with his or her overall preference in terms of order. His or her top choice should naturally be the one that they want to see win the election. By casting a vote in which the ordering of the non-preferred candidates are listed can change the outcome in favor of the preferred candidate. A voting system is **manipulable** if there exists *at least one* way a voter can achieve a preferred outcome by changing his or her preference ballot.

⚿ Key idea

The Borda count method is subject to manipulation under certain conditions. One of these conditions is having three voters and four candidates. Note: Other conditions will be discussed later.

↷ Example A

Consider the following election with four candidates and five voters.

Election 1

Rank	Number of voters (5)				
	1	1	1	1	1
First	B	A	A	B	D
Second	A	B	B	A	C
Third	D	C	D	C	A
Fourth	C	D	C	D	B

Show that if the Borda count is being used, the voter on the left can manipulate the outcome (assuming the above ballot represents his true preferences).

Solution

Preference	1st place votes × 3	2nd place votes × 2	3rd place votes × 1	4th place votes × 0	Borda score
A	2×3	2×2	1×1	0×0	11
B	2×3	2×2	0×1	1×0	10
C	0×3	1×2	2×1	2×0	4
D	1×3	0×2	2×1	2×0	5

With the given ballots, the winner using the Borda count is *A*. However, if the leftmost voter changes his or her preference ballot, we have the following.

Election 2

Rank	Number of voters (5)				
	1	1	1	1	1
First	B	A	A	B	D
Second	C	B	B	A	C
Third	D	C	D	C	A
Fourth	A	D	C	D	B

Continued on next page

continued

Preference	1ˢᵗ place votes × 3	2ⁿᵈ place votes × 2	3ʳᵈ place votes × 1	4ᵗʰ place votes × 0	Borda score
A	2 × 3	1 × 2	1 × 1	1 × 0	9
B	2 × 3	2 × 2	0 × 1	1 × 0	10
C	0 × 3	2 × 2	2 × 1	1 × 0	6
D	1 × 3	0 × 2	2 × 1	2 × 0	5

With the new ballots, the winner using the Borda count is B.

⌬ Key idea

The term **unilateral change** is used when one voter (as opposed to a group of voters) changes his or her ballot.

⌬ Key idea

Definition of Manipulability: A voting system is said to be **manipulable** if there exist two sequences of preference list ballots and a voter (call the voter j) such that
- Neither election results in a tie. (Ties in an election present a problem in determining sincere preference.)
- The only ballot change is by voter j (This is a unilateral change)
- Voter j prefers the outcome (overall winner) of the second election even though the first election showed his or her true (overall order) preferences.

Section 10.1   Majority Rule and Condorcet's Method

⌬ Key idea

In this section, like in Chapter 9, it is assumed that the number of voters is odd.

⌬ Key idea

(Restated from Chapter 9) When there are only two candidates or alternatives, May's theorem states that **majority rule** is the only voting method that satisfies three desirable properties, given an odd number of voters and no ties. The three properties satisfied by majority rule are:
1. All voters are treated equally.
2. Both candidates are treated equally.
3. If a single voter who voted for the loser, B, changes his mind and votes for the winner, A, then A is still the winner. This is what is called **monotone.**

Because in the two-candidate case, there are only two possible rankings (A over B or B over A), the monotonic property of majority rule is equivalent to the non-manipulability of this voting system, given the voter and candidate restriction.

⌬ Key idea

Condorcet's method is non-manipulable by a unilateral change in vote. This statement **does not** consider the possibility that an election manipulation could result in no winner. It is possible to go from having a winner to having no winner by unilateral change in vote. If this is a desired outcome by the disingenuous voter, then Condorcet's method can be altered by a unilateral change in vote.

⌘ Example B

Consider the following election with four candidates and three voters.

Election 1

	Number of voters (3)		
Rank	**1**	**1**	**1**
First	C	A	B
Second	A	C	A
Third	D	D	D
Fourth	B	B	C

Show that if Condorcet's method is being used, the voter on the left can change the outcome so that there is no winner.

Solution

There are 6 one-on-one contests as summarized below.

A vs B	A:	2	B:	1
A vs C	A:	2	C:	1
A vs D	A:	3	D:	0
B vs C	B:	1	C:	2
B vs D	B:	1	D:	2
C vs D	C:	2	D:	1

Since A can beat the other candidates in a one-on-one contest, A is declared the winner by Condorcet's method.

Election 2

	Number of voters (3)		
Rank	**1**	**1**	**1**
First	C	A	B
Second	B	C	A
Third	A	D	D
Fourth	D	B	C

A vs B	A:	1	B:	2
A vs C	A:	2	C:	1
A vs D	A:	3	D:	0
B vs C	B:	1	C:	2
B vs D	B:	2	D:	1
C vs D	C:	2	D:	1

Since no candidate can beat all other candidates in a one-on-one contest, there is no winner by Condorcet's method.

Section 10.2   Other Voting Systems for Three of More Candidates

⛬ Key idea

The Borda count method is non-manipulable for three candidates, regardless of the number of voters.

⛬ Key idea

The Borda count method is manipulable for four or more candidates (and two or more voters).

ᕫ Example C

Consider the following election with four candidates and two voters.

Election 1

	Number of voters (2)	
Rank	1	1
First	A	C
Second	C	B
Third	B	A
Fourth	D	D

Show that if the Borda count is being used, the voter on the left can manipulate the outcome (assuming the above ballot represents his true preferences).

Solution

Preference	1st place votes × 3	2nd place votes × 2	3rd place votes × 1	4th place votes × 0	Borda score
A	1×3	0×2	1×1	0×0	4
B	0×3	1×2	1×1	0×0	3
C	1×3	1×2	0×1	0×0	5
D	0×3	0×2	0×1	2×0	0

With the given ballots, the winner using the Borda count is C. However, if the left-most voter changes his or her preference ballot, we have the following.

Election 2

	Number of voters (2)	
Rank	1	1
First	A	C
Second	D	B
Third	B	A
Fourth	C	D

Preference	1st place votes × 3	2nd place votes × 2	3rd place votes × 1	4th place votes × 0	Borda score
A	1×3	0×2	1×1	0×0	4
B	0×3	1×2	1×1	0×0	3
C	1×3	0×2	0×1	1×0	3
D	0×3	1×2	0×1	1×0	2

With the new ballots, the winner using the Borda count is A.

✏️ Question 1

Consider Example 2 from the text. Is it possible to use the preference list ballots from Example C (last page) to create an example of manipulating the Borda count with five candidates and six voters? Justify your yes/no response.

Answer

Yes.

🔑 Key idea

The plurality runoff rule is manipulable.

᧬ Example D

Consider the following election with four candidates and five voters.

Election 1

	Number of voters (5)				
Rank	1	1	1	1	1
First	D	C	C	B	D
Second	B	B	B	A	B
Third	C	A	A	C	A
Fourth	A	D	D	D	C

Show how the left-most voter can secure a **more preferred** outcome by a unilateral change of ballot using the plurality runoff rule.

Solution

Since C and D have the most number of first-place votes, A and B are eliminated.

	Number of voters (5)				
Rank	1	1	1	1	1
First	D	C	C	C	D
Second	C	D	D	D	C

Since C has the most number of first-place votes, the winner using the plurality runoff rule is C. But the winner becomes B if the leftmost voter changes his or her ballot as the following shows.

Election 2

	Number of voters (5)				
Rank	1	1	1	1	1
First	B	C	C	B	D
Second	D	B	B	A	B
Third	C	A	A	C	A
Fourth	A	D	D	D	C

Since B and C have the most number of first-place votes, A and D are eliminated.

	Number of voters (5)				
Rank	1	1	1	1	1
First	B	C	C	B	B
Second	C	B	B	C	C

Since B has the most number of first-place votes, the winner using the plurality runoff rule is B. For the first voter, having B win the election was **more preferred** than having C win the election.

⊶ Key idea

The Hare system is manipulable.

ᨳ Example E

Election 1

	Number of voters (5)				
Rank	**1**	**1**	**1**	**1**	**1**
First	D	C	C	B	D
Second	B	B	B	A	B
Third	C	A	A	C	A
Fourth	A	D	D	D	C

Show how the left-most voter can secure a **more preferred** outcome by a unilateral change of ballot using the Hare system.

Solution

A has the fewest first-place votes and is thus eliminated.

	Number of voters (5)				
Rank	**1**	**1**	**1**	**1**	**1**
First	D	C	C	B	D
Second	B	B	B	C	B
Third	C	D	D	D	C

B now has the fewest first-place votes and is eliminated

	Number of voters (5)				
Rank	**1**	**1**	**1**	**1**	**1**
First	D	C	C	C	D
Second	C	D	D	D	C

D now has the fewest first-place votes and is eliminated, leaving C as the winner.

Election 2

	Number of voters (5)				
Rank	**1**	**1**	**1**	**1**	**1**
First	B	C	C	B	D
Second	D	B	B	A	B
Third	C	A	A	C	A
Fourth	A	D	D	D	C

A has the fewest first-place votes and is eliminated.

	Number of voters (5)				
Rank	**1**	**1**	**1**	**1**	**1**
First	B	C	C	B	D
Second	D	B	B	C	B
Third	C	D	D	D	C

D now has the fewest first-place votes and is eliminated

	Number of voters (5)				
Rank	**1**	**1**	**1**	**1**	**1**
First	B	C	C	B	B
Second	C	B	B	C	C

C now has the fewest first-place votes and is eliminated, leaving B as the winner. For the first voter, having B win the election was **more preferred** than having C win the election.

⊶ **Key idea**

Sequential pairwise voting, with agenda, is manipulable by having the agenda altered.

⬭ **Example F**

Consider the following election with four candidates and three voters.

	Number of voters (3)		
Rank	**1**	**1**	**1**
First	A	B	D
Second	B	C	C
Third	C	D	A
Fourth	D	A	B

Show that sequential pairwise voting, with agenda A, B, C, D, can be manipulated by the voter on the left by a change of agenda. (assuming the above ballot represents his true preferences).

Solution

Looking at the 6 one-on-one contests we can more readily see the solution.

A vs B	A:	2	B:	1
A vs C	A:	1	C:	2
A vs D	A:	1	D:	2
B vs C	B:	2	C:	1
B vs D	B:	2	D:	1
C vs D	C:	2	D:	1

In sequential pairwise voting with the agenda A, B, C, D, we first pit A against B. Thus, A wins by a score of 2 to 1. A moves on to confront C. C wins by a score of 2 to 1. C moves on to confront D. C wins by a score of 2 to 1. Thus, C is the winner by sequential pairwise voting with the agenda A, B, C, D.

If the voter on the left changes the agenda to B, C, D, A, we have the following.

We first pit B against C. Thus, B wins by a score of 2 to 1. B moves on to confront D. B wins by a score of 2 to 1. B moves on to confront A. A wins by a score of 2 to 1. Thus, A is the winner by sequential pairwise voting with the agenda B, C, D, A.

✐ **Question 2**

Consider the following election with four candidates and 3 voters.

	Number of voters (3)		
Rank	**1**	**1**	**1**
First	A	B	D
Second	B	C	A
Third	C	D	B
Fourth	D	A	C

If sequential pairwise voting, with agenda is used, is it possible to make all candidates winners (i.e. four separate manipulations/agendas) by different agendas? Explain your yes/no answer.

Answer

Yes.

⌐ Key idea

Plurality voting *can be* **group-manipulable**. Group-manipulable is when a group of voters can change the outcome of an election (as a group) to something they all prefer.

∽ Example G

Consider the following election with four candidates and 11 voters.

Election 1

Rank	Number of voters (11)		
	2	4	5
First	B	C	D
Second	C	B	A
Third	A	D	C
Fourth	D	A	B

Show that if plurality voting is used, the group of voters on the left can secure a **more preferred** outcome.

Solution

Since Candidate D has the most first-place votes, D is declared the winner.

Election 2

Rank	Number of voters (11)		
	2	4	5
First	C	C	D
Second	B	B	A
Third	A	D	C
Fourth	D	A	B

Since the group on the left changed their ballots, C now has 6 (the most votes) and is declared the winner. Having C win the election was **more preferred** by the left most group of voters, rather than having D win the election.

✐ Question 3

Consider the following election with four candidates and 11 voters.

Rank	Number of voters (11)		
	2	5	4
First	B	C	D
Second	C	B	A
Third	A	D	C
Fourth	D	A	B

If plurality voting is used, can the group of voters on the left secure a **more preferred** outcome? Explain your yes/no answer.

Answer

No.

Section 10.3   Impossibility

⌐ Key idea

Condorcet's method has very desirable properties including the following four.

- Elections never result in ties.
- It satisfies the Pareto condition. (It states that if everyone prefers one candidate, say A, to another, say B, then B cannot be the winner.)
- It is non-manipulable. (In a dictatorship all ballots except that of the dictator are ignored.)

A less than desirable outcome though is that Condorcet's method could produce no winner at all.

⌐ Key idea

An important theorem in social choice is the **Gibbard-Satterthwaite Theorem** ("GS theorem" for short). It says that with three or more candidates and any number of voters, there does not exist (and never will exist) a voting system that always has all of the following features.

- a winner
- no ties
- satisfies the Pareto condition
- non-manipulable
- not a dictatorship.

⌐ Key idea

A *weak* version of the Gibbard-Satterthwaite Theorem refers to any voting system for three candidates that agrees with Condorcet's method whenever there is a Condorcet winner. This voting system must also produce a unique winner when confronted by the ballots in the Condorcet voting paradox. Given these conditions, this voting system is manipulable.

Section 10.4   The Chair's Paradox

⌐ Key idea

A (single) choice of which candidate to vote for will be called a **strategy**. If a voter is rational, he or she will not vote for their least-preferred candidate. In the text example of the chair's paradox, it is assumed that there are three candidates and three voters. If a candidate gets two or three votes, he or she wins. If each candidate gets one vote (three-way tie), then the chair has **tie-breaking** power as his or her candidate is the winner. The paradox that occurs in this voting set-up is although the chair has tie-breaking power, the eventual winner (given the different voting strategies) is his or her least-preferred candidate.

⌐ Key idea

The strategy of choosing a candidate, say X, **weakly dominates** another choice, say Y, if the choice of X yields outcomes that are either the same or better than the choice of Y.

⚷ **Key idea**

To examine the text example of the chair's paradox, consider the names Adam, Nadia, and Zeki as candidates. The voters are Scott (Chair), Dan, and Sami. Although each voter can cast only one vote for one candidate, they each do have overall preferences as follows.

Rank	Number of voters (3)		
	Scott	**Dan**	**Sami**
First	Adam	Nadia	Zeki
Second	Nadia	Zeki	Adam
Third	Zeki	Adam	Nadia

Now for Scott (Chair), voting for Adam weakly dominates voting for Nadia. The possible outcomes if Scott votes for Adam are as follows.

Scott	Dan	Sami
Adam	Nadia	Zeki

Adam wins because Scott (Chair) breaks the tie.

Scott	Dan	Sami
Adam	Nadia	Adam

Adam wins because of two-thirds vote.

Scott	Dan	Sami
Adam	Zeki	Zeki

Zeki wins because of two-thirds vote.

Scott	Dan	Sami
Adam	Zeki	Adam

Adam wins because of two-thirds vote.

The possible outcomes if Scott votes for Nadia are as follows.

Scott	Dan	Sami
Nadia	Nadia	Zeki

Nadia wins because of two-thirds vote.

Scott	Dan	Sami
Nadia	Nadia	Adam

Nadia wins because of two-thirds vote.

Scott	Dan	Sami
Nadia	Zeki	Zeki

Zeki wins because of two-thirds vote.

Scott	Dan	Sami
Nadia	Zeki	Adam

Nadia wins because Scott (Chair) breaks the tie.

Clearly for Scott (Chair) the choice of Adam yields more desirable results. Since it is assumed that Scott is rational, we know that Adam will be Scott's choice.

Now, examining Dan's options. (The text examines the case of Sami, *C*, followed by Dan, *B*.)

Scott	Dan	Sami
Adam	Nadia	Zeki

Adam wins because Scott (Chair) breaks the tie.

Scott	Dan	Sami
Adam	Nadia	Adam

Adam wins because of two-thirds vote.

Scott	Dan	Sami
Adam	Zeki	Zeki

Zeki wins because of two-thirds vote.

Scott	Dan	Sami
Adam	Zeki	Adam

Adam wins because of two-thirds vote.

Since an outcome of Zeki is more favorable to Dan, voting for Zeki weakly dominates voting for Nadia. Unfortunately, Dan's top choice of Nadia is not possible. This leaves Sami's choices to be examined.

Scott	Dan	Sami
Adam	Zeki	Zeki

Zeki wins because of two-thirds vote.

Scott	Dan	Sami
Adam	Zeki	Adam

Adam wins because of two-thirds vote.

Since Zeki is the preferred choice of Sami, voting for Zeki weakly dominates voting for Adam. So the winner would be Zeki, which is the least-preferred choice of Scott (Chair).

Homework Help

Exercises 1 – 3

Carefully read Section 10.1 before responding to these exercises. For each exercise start off by setting up Election 1 that produces a candidate, say B, as the winner given the voting method. Taking the ballots in the first election to be the sincere preferences of the voters, then change a ballot (one that prefers A to B) to secure a more favorable outcome by the submission of a disingenuous ballot. The following tables may be helpful in setting up the two elections.

Election 1

Rank	Number of voters (3)
First	
Second	

Election 2

Rank	Number of voters (3)
First	
Second	

Exercise 4 – 5

Carefully read Section 10.1 before responding to these exercises. Pay special attention to May's theorem. Example of voting systems should not be complicated.

Exercise 6

Review Condorcet's method and consider the three one-on-one scores of D versus H, D versus J, and H versus J.

Exercise 7

Given the preference list ballots, determine the winner by the Borda count voting method.

Election 1

	Number of voters (2)	
Rank	1	1
First	B	A
Second	C	D
Third	A	C
Fourth	D	B

Preference	1st place votes × 3	2nd place votes × 2	3rd place votes × 1	4th place votes × 0	Borda score
A	×3	×2	×1	×0	
B	×3	×2	×1	×0	
C	×3	×2	×1	×0	
D	×3	×2	×1	×0	

Change the leftmost voter preference ballot to manipulate the election.

Election 2

	Number of voters (2)	
Rank	1	1
First		A
Second		D
Third		C
Fourth		B

Preference	1st place votes × 3	2nd place votes × 2	3rd place votes × 1	4th place votes × 0	Borda score
A	×3	×2	×1	×0	
B	×3	×2	×1	×0	
C	×3	×2	×1	×0	
D	×3	×2	×1	×0	

Exercise 8

One way to get an example of manipulation of the Borda count with seven candidates and eight voters is to alter the elections in Example 2 of the text by adding F and G to the bottom of each of the six ballots in both elections, and then adding in the two rightmost columns. One could also add two ballots canceling each other out first, and then add F and G to the bottom of all eight ballots.

Election 1

Rank	Number of voters (8)							
	1	1	1	1	1	1	1	1
First	A	B	A	E	A	E		
Second	B	C	B	D	B	D		
Third	C	A	C	C	C	C		
Fourth	D	D	D	B	D	B		
Fifth	E	E	E	A	E	A		
Sixth								
Seventh								

Preference	1st place votes × 6	2nd place votes × 5	3rd place votes × 4	4th place votes × 3	5th place votes × 2	6th place votes × 1	7th place votes × 0	Borda score
A	×6	×5	×4	×3	×2	×1	×0	
B	×6	×5	×4	×3	×2	×1	×0	
C	×6	×5	×4	×3	×2	×1	×0	
D	×6	×5	×4	×3	×2	×1	×0	
E	×6	×5	×4	×3	×2	×1	×0	
F	×6	×5	×4	×3	×2	×1	×0	
G	×6	×5	×4	×3	×2	×1	×0	

Election 2

Rank	Number of voters (8)							
	1	1	1	1	1	1	1	1
First		B	A	E	A	E		
Second		C	B	D	B	D		
Third		A	C	C	C	C		
Fourth		D	D	B	D	B		
Fifth		E	E	A	E	A		
Sixth								
Seventh								

Preference	1st place votes × 6	2nd place votes × 5	3rd place votes × 4	4th place votes × 3	5th place votes × 2	6th place votes × 1	7th place votes × 0	Borda score
A	×6	×5	×4	×3	×2	×1	×0	
B	×6	×5	×4	×3	×2	×1	×0	
C	×6	×5	×4	×3	×2	×1	×0	
D	×6	×5	×4	×3	×2	×1	×0	
E	×6	×5	×4	×3	×2	×1	×0	
F	×6	×5	×4	×3	×2	×1	×0	
G	×6	×5	×4	×3	×2	×1	×0	

Exercise 9
Election 1

	Number of voters (3)		
Rank	1	1	1
First	A	B	B
Second	B	A	A
Third	C	C	C
Fourth	D	D	D

Preference	1st place votes × 3	2nd place votes × 2	3rd place votes × 1	4th place votes × 0	Borda score
A	× 3	× 2	× 1	× 0	
B	× 3	× 2	× 1	× 0	
C	× 3	× 2	× 1	× 0	
D	× 3	× 2	× 1	× 0	

Election 2

	Number of voters (3)		
Rank	1	1	1
First		B	B
Second		A	A
Third		C	C
Fourth		D	D

Preference	1st place votes × 3	2nd place votes × 2	3rd place votes × 1	4th place votes × 0	Borda score
A	× 3	× 2	× 1	× 0	
B	× 3	× 2	× 1	× 0	
C	× 3	× 2	× 1	× 0	
D	× 3	× 2	× 1	× 0	

Exercise 10
Election 1

	Number of voters (5)				
Rank	1	1	1	1	1
First	A	B	B		
Second	B	A	A		
Third	C	C	C		
Fourth	D	D	D		

Preference	1st place votes × 3	2nd place votes × 2	3rd place votes × 1	4th place votes × 0	Borda score
A	× 3	× 2	× 1	× 0	
B	× 3	× 2	× 1	× 0	
C	× 3	× 2	× 1	× 0	
D	× 3	× 2	× 1	× 0	

Continued on next page

Exercise 10 continued
Election 2

	Number of voters (5)				
Rank	1	1	1	1	1
First		B	B		
Second		A	A		
Third		C	C		
Fourth		D	D		

Preference	1st place votes × 3	2nd place votes × 2	3rd place votes × 1	4th place votes × 0	Borda score
A	× 3	× 2	× 1	× 0	
B	× 3	× 2	× 1	× 0	
C	× 3	× 2	× 1	× 0	
D	× 3	× 2	× 1	× 0	

Exercise 11
Election 1

	Number of voters (9)								
Rank	1	1	1	1	1	1	1	1	1
First	A	B	B						
Second	B	A	A						
Third	C	C	C						
Fourth	D	D	D						
Fifth									
Sixth									

Preference	1st place votes × 5	2nd place votes × 4	3rd place votes × 3	4th place votes × 2	5th place votes × 1	6th place votes × 0	Borda score
A	×5	×4	×3	×2	×1	×0	
B	×5	×4	×3	×2	×1	×0	
C	×5	×4	×3	×2	×1	×0	
D	×5	×4	×3	×2	×1	×0	
E	×5	×4	×3	×2	×1	×0	
F	×5	×4	×3	×2	×1	×0	

Election 2

	Number of voters (9)								
Rank	1	1	1	1	1	1	1	1	1
First		B	B						
Second		A	A						
Third		C	C						
Fourth		D	D						
Fifth									
Sixth									

Preference	1st place votes × 5	2nd place votes × 4	3rd place votes × 3	4th place votes × 2	5th place votes × 1	6th place votes × 0	Borda score
A	×5	×4	×3	×2	×1	×0	
B	×5	×4	×3	×2	×1	×0	
C	×5	×4	×3	×2	×1	×0	
D	×5	×4	×3	×2	×1	×0	
E	×5	×4	×3	×2	×1	×0	
F	×5	×4	×3	×2	×1	×0	

Exercise 12
Election 1

	Number of voters (4)			
Rank	**1**	**1**	**1**	**1**
First	*B*	*D*	*C*	*B*
Second	*C*	*C*	*A*	*A*
Third	*D*	*A*	*B*	*C*
Fourth	*A*	*B*	*D*	*D*

Preference	1st place votes × 3	2nd place votes × 2	3rd place votes × 1	4th place votes × 0	Borda score
A	× 3	× 2	× 1	× 0	
B	× 3	× 2	× 1	× 0	
C	× 3	× 2	× 1	× 0	
D	× 3	× 2	× 1	× 0	

Election 2

	Number of voters (4)			
Rank	**1**	**1**	**1**	**1**
First		*D*	*C*	*B*
Second		*C*	*A*	*A*
Third		*A*	*B*	*C*
Fourth		*B*	*D*	*D*

Preference	1st place votes × 3	2nd place votes × 2	3rd place votes × 1	4th place votes × 0	Borda score
A	× 3	× 2	× 1	× 0	
B	× 3	× 2	× 1	× 0	
C	× 3	× 2	× 1	× 0	
D	× 3	× 2	× 1	× 0	

Exercise 13
Election 1

	Number of voters (4)			
Rank	**1**	**1**	**1**	**1**
First	*A*	*C*	*B*	*D*
Second	*B*	*A*	*D*	*C*
Third	*C*	*B*	*C*	*A*
Fourth	*D*	*D*	*A*	*B*

Preference	1st place votes × 3	2nd place votes × 2	3rd place votes × 1	4th place votes × 0	Borda score
A	× 3	× 2	× 1	× 0	
B	× 3	× 2	× 1	× 0	
C	× 3	× 2	× 1	× 0	
D	× 3	× 2	× 1	× 0	

Continued on next page

Exercise 13 continued

Election 2

	Number of voters (4)			
Rank	**1**	**1**	**1**	**1**
First		C	B	D
Second		A	D	C
Third		B	C	A
Fourth		D	A	B

Preference	1st place votes × 3	2nd place votes × 2	3rd place votes × 1	4th place votes × 0	Borda score
A	× 3	× 2	× 1	× 0	
B	× 3	× 2	× 1	× 0	
C	× 3	× 2	× 1	× 0	
D	× 3	× 2	× 1	× 0	

Exercise 14

In this exercise, award 1 point to the winner of the one-on-one competition and 0 to the loser. If it is a tie, award ½ point to each.

Election 1

	Number of voters (4)			
Rank	**1**	**1**	**1**	**1**
First	A	C	A	D
Second	B	E	E	B
Third	C	D	D	E
Fourth	D	B	C	C
Fifth	E	A	B	A

There are 10 one-to-one contests. Ties are possible since we have an even number of voters.

A versus B: _____   B versus D: _____

A versus C: _____   B versus E: _____

A versus D: _____   C versus D: _____

A versus E: _____   C versus E: _____

B versus C: _____   D versus E: _____

You may find it helpful to summarize your results in the following table.

	A	B	C	D	E
Total					

Continued on next page

Exercise 14 continued
Election 2

	Number of voters (4)			
Rank	**1**	**1**	**1**	**1**
First		C	A	D
Second		E	E	B
Third		D	D	E
Fourth		B	C	C
Fifth		A	B	A

A versus B: _____ B versus D: _____

A versus C: _____ B versus E: _____

A versus D: _____ C versus D: _____

A versus E: _____ C versus E: _____

B versus C: _____ D versus E: _____

You may find it helpful to summarize your results in the following table.

	A	B	C	D	E
Total					

Exercise 15
Election 1

	Number of voters (5)				
Rank	**1**	**1**	**1**	**1**	**1**
First	A	B	B	A	A
Second	B	C	C	C	C
Third	C	A	A	B	B

Election 2

	Number of voters (5)				
Rank	**1**	**1**	**1**	**1**	**1**
First		B	B	A	A
Second		C	C	C	C
Third		A	A	B	B

	Number of voters (5)				
Rank	**1**	**1**	**1**	**1**	**1**
First					
Second					

Exercise 16
Review the Hare voting system before starting this exercise.
Election 1

	Number of voters (5)				
Rank	**1**	**1**	**1**	**1**	**1**
First	*A*	*B*	*C*	*C*	*D*
Second	*B*	*A*	*B*	*B*	*B*
Third	*C*	*C*	*A*	*A*	*C*
Fourth	*D*	*D*	*D*	*D*	*A*

Election 2

	Number of voters (5)				
Rank	**1**	**1**	**1**	**1**	**1**
First		*B*	*C*	*C*	*D*
Second		*A*	*B*	*B*	*B*
Third		*C*	*A*	*A*	*C*
Fourth		*D*	*D*	*D*	*A*

	Number of voters (5)				
Rank	**1**	**1**	**1**	**1**	**1**
First					
Second					
Third					

	Number of voters (5)				
Rank	**1**	**1**	**1**	**1**	**1**
First					
Second					

Exercise 17
Review the plurality runoff rule before starting this exercise.
Election 1

	Number of voters (5)				
Rank	**1**	**1**	**1**	**1**	**1**
First	*A*	*A*	*C*	*C*	*B*
Second	*B*	*B*	*A*	*A*	*C*
Third	*C*	*C*	*B*	*B*	*A*

	Number of voters (5)				
Rank	**1**	**1**	**1**	**1**	**1**
First					
Second					

Election 2

	Number of voters (5)				
Rank	**1**	**1**	**1**	**1**	**1**
First		*A*	*C*	*C*	*B*
Second		*B*	*A*	*A*	*C*
Third		*C*	*B*	*B*	*A*

	Number of voters (5)				
Rank	**1**	**1**	**1**	**1**	**1**
First					
Second					

Exercise 18
Review sequential pairwise, with agenda, voting method before starting this exercise.
Election 1

	Number of voters (3)		
Rank	1	1	1
First	A	B	C
Second	B	C	A
Third	C	A	B

Election 2

	Number of voters (3)		
Rank	1	1	1
First		B	C
Second		C	A
Third		A	B

Exercise 19

	Number of voters (3)		
Rank	1	1	1
First	A	C	B
Second	B	A	D
Third	D	B	C
Fourth	C	D	A

There are 12 different possible agendas to consider with four candidates.

A, B, C, D (equivalent to B, A, C, D)   B, C, A, D (equivalent to C, B, A, D)

A, B, D, C (equivalent to B, A, D, C)   B, C, D, A (equivalent to C, B, D, A)

A, C, B, D (equivalent to C, A, B, D)   B, D, A, C (equivalent to D, B, A, C)

A, C, D, B (equivalent to C, A, B, D)   B, D, C, A (equivalent to D, B, C, A)

A, D, B, C (equivalent to D, A, B, C)   C, D, A, B (equivalent to D, C, A, B)

A, D, C, B (equivalent to D, A, C, B)   C, D, B, A (equivalent to D, C, B, A)

Exercise 20
Review the Pareto condition before starting this exercise. Look under the first ☞ Key idea from Section 10.3 in this Study Guide.

Exercise 21
Review the plurality rule before starting this exercise.
Election 1

22%	23%	15%	29%	7%	4%
D	D	H	H	J	J
H	J	D	J	H	D
J	H	J	D	D	H

Consider what would happen if the voters in the 7% group all change their ballots.
Election 2

22%	23%	15%	29%	7%	4%
D	D	H	H		J
H	J	D	J		D
J	H	J	D		H

Exercise 22

(a) Assume that the winner with the voting paradox ballots is A. Consider the following two elections:

Election 1

Rank	Number of voters (3)		
First	A	B	C
Second	B	C	A
Third	C	A	B

Election 2

Rank	Number of voters (3)	
First	A	C
Second	B	A
Third	C	B

In Election 1, the winner is A (our assumption in this case) and in Election 2, the winner is C (because we are assuming that our voting system agrees with Condorcet's method when there is a Condorcet winner, as C is here).

(b) Assume that the winner with the voting paradox ballots is B. Consider the following two elections:

Election 1

Rank	Number of voters (3)		
First	A	B	C
Second	B	C	A
Third	C	A	B

Election 2

Rank	Number of voters (3)	
First	A	B
Second	B	C
Third	C	A

In Election 1, the winner is B (our assumption in this case) and in Election 2, the winner is A (because we are assuming that our voting system agrees with Condorcet's method when there is a Condorcet winner, as A is here).

Exercise 23

Review what a dictator is before starting this exercise. Look under the first ↝ Key idea from Section 10.3 in this Study Guide.

Exercises 24 – 25

Carefully read the conditions being used as a voting rules.

Exercises 26 – 27

Carefully read in Section 10.4 what it means to *weakly dominate*. Reading through the example in the text along with the similar version in this Study Guide should help you in coming up with the scenario to show that voting for a certain candidate does not weakly dominate your strategy of voting for another.

Do You Know the Terms?

Cut out the following 11 flashcards to test yourself on Review Vocabulary. You can also find these flashcards at http://www.whfreeman.com/fapp7e.

Chapter 10 **The Manipulability of Voting Systems** **Agenda manipulation**	**Chapter 10** **The Manipulability of Voting Systems** **Chair's paradox**
Chapter 10 **The Manipulability of Voting Systems** **Disingenuous ballot**	**Chapter 10** **The Manipulability of Voting Systems** **Gibbard-Satterthwaite theorem**
Chapter 10 **The Manipulability of Voting Systems** **Group manipulability**	**Chapter 10** **The Manipulability of Voting Systems** **Manipulation**
Chapter 10 **The Manipulability of Voting Systems** **May's theorem for manipulability**	**Chapter 10** **The Manipulability of Voting Systems** **Strategy**

The fact that with three voters and three candidates, the voter with tie-breaking power (the "chair") can — if all three voters act rationally in their own self-interest — end up with his least-preferred candidate as the election winner.

The ability to control who wins an election with sequential pairwise voting by a choice of the agenda — that is, a choice of the order in which the one-on-one contests will be held.

Alan Gibbard's and Mark Satterthwaite's independent discovery that every voting system for three or more candidates and any number of voters that satisfies the Pareto condition, always produces a unique winner, and is not a dictatorship, can be manipulated.

Any ballot that does not represent a voter's true preferences.

A voting system is manipulable if there exists at least one election in which a voter can change his or her ballot (with the ballots of all other voters left unchanged) in such a way that he or she prefers the winner of the new election to the winner of the old election, assuming that the original ballots represent the true preferences of the voters.

A voting system is group manipulable if there exists at least one election in which a group of voters can change their ballots (with the ballots of voters not in the group left unchanged) in such a way that they all prefer the winner of the new election to the winner of the old election, assuming that the original ballots represent the true preferences of these voters.

In the chair's paradox, a choice of which candidate to vote for is called a strategy. This is a special case of the use of the term in general game-theoretic situations.

Kenneth May's discovery that for two candidates and an odd number of voters, majority rule is the only voting system that treats both candidates equally, all voters equally, and is non-manipulable.

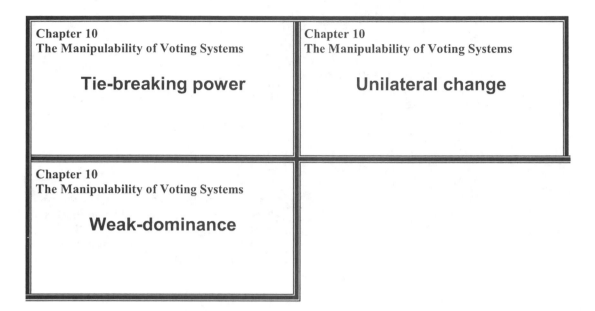

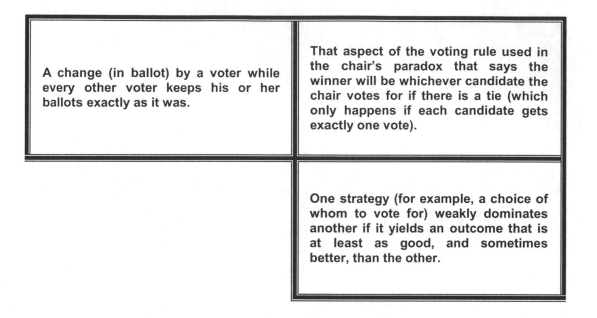

A change (in ballot) by a voter while every other voter keeps his or her ballots exactly as it was.

That aspect of the voting rule used in the chair's paradox that says the winner will be whichever candidate the chair votes for if there is a tie (which only happens if each candidate gets exactly one vote).

One strategy (for example, a choice of whom to vote for) weakly dominates another if it yields an outcome that is at least as good, and sometimes better, than the other.

Practice Quiz

1. Which of the following is *not* part of the conditions that define manipulability?

   a. The voter that changes his or her preference list ballot must manipulate the election so that his or her top choice is declared the winner.

   b. The voter that changes his or her preference list ballot must force a tie in the second election.

   c. The voter that changes his or her preference list ballot prefers the outcome of the second election.

2. "Among all two-candidate voting systems that never result in a tie, majority rule is the only one that treats all voters equally, both candidates equally, and is non-manipulable." This statement is known as

   a. the chair's paradox.

   b. May's theorem.

   c. The Gibbard-Satterthwaite theorem.

3. Consider the following election with four candidates and three voters. If the Borda count method is used, the winner of the first election is B. What can be said about the left-most voter?

   Election 1

Rank	Number of voters (3)		
	1	1	1
First	A	B	C
Second	D	C	B
Third	B	A	A
Fourth	C	D	D

Borda Scores	
A	5
B	6
C	5
D	2

   a. In Election 2, the left-most voter cannot manipulate his or her ballot for a more favorable outcome.

   b. In Election 2, the left-most voter can manipulate his or her ballot as follows.

   Election 2

Rank	Number of voters (3)		
	1	1	1
First	A	B	C
Second	C	C	B
Third	D	A	A
Fourth	B	D	D

Borda Scores	
A	5
B	5
C	7
D	1

   c. The left-most voter can manipulate the vote if the right-most voter will also change his or her preference ballot.

4. Consider the following election with four candidates and three voters. Assume that these represent true preferences and sequential pairwise voting, with agenda A, B, C, D.

Rank	Number of voters (3)		
	1	1	1
First	A	B	C
Second	D	C	D
Third	B	D	B
Fourth	C	A	A

Which of the following agendas will allow the voter on the right to manipulate the outcome to a *more* favorable one?

a. A, C, D, B

b. D, B, C, A

c. C, B, A, D

5. Consider the following election with four candidates and three voters.

Election 1

Rank	Number of voters (3)		
	1	1	1
First	B	C	D
Second	C	B	B
Third	D	D	C
Fourth	A	A	A

Which of the following Election 2 preference list ballots demonstrates that if Condorcet's method is being used, the voter on the right can change the outcome from having a winner to having no winner.

a. Election 2

Rank	Number of voters (3)		
	1	1	1
First	B	C	D
Second	C	B	B
Third	D	D	A
Fourth	A	A	C

b. Election 2

Rank	Number of voters (3)		
	1	1	1
First	B	C	D
Second	C	B	A
Third	D	D	B
Fourth	A	A	C

c. Election 2

Rank	Number of voters (3)		
	1	1	1
First	B	C	D
Second	C	B	C
Third	D	D	B
Fourth	A	A	A

6. Which of the following voting systems are not manipulable by a single voter but is by a group of voters?

   a. Plurality voting

   b. Borda count

   c. Hare system

7. Consider the following election with three candidates and five voters.

   Election 1

Rank	Number of voters (5)				
	1	1	1	1	1
First	B	C	B	A	C
Second	A	A	C	C	A
Third	C	B	A	B	B

   Assume that these represent true preferences and the Hare voting method is used. Which of the following is true regarding the left-most voter.

   a. He or she can obtain a more favorable outcome by the following unilateral change.

      Election 2

Rank	Number of voters (5)				
	1	1	1	1	1
First	B	C	B	A	C
Second	C	A	C	C	A
Third	A	B	A	B	B

   b. He or she can obtain a more favorable outcome by the following unilateral change.

      Election 2

Rank	Number of voters (5)				
	1	1	1	1	1
First	A	C	B	A	C
Second	B	A	C	C	A
Third	C	B	A	B	B

   c. He or she cannot obtain a more favorable outcome by a unilateral change

8. The chair's paradox refers to

   a. the manipulability of an election by the chair of a committee.

   b. the chair obtaining a more favorable outcome by actually handing over tie-breaking power to one of the other voters.

   c. having an election with three voters and three candidates.

9. As discussed in the chair's paradox, voting for a candidate, say X, that *weakly dominates* another, say Y, refers to

   a. X having more potential to win the election over Y.

   b. choosing X over Y would yield a better outcome for the voter.

   c. choosing X over Y would yield a better or the same outcome for the voter.

10. Consider the following election with four candidates and 13 voters.

Election 1

Rank	Number of voters (13)		
	3	**4**	**6**
First	B	C	D
Second	A	B	A
Third	C	A	C
Fourth	D	D	B

If plurality voting is used, which of the following demonstrates that a group of voters secured a more preferred outcome?

a. Election 2

Rank	Number of voters (13)		
	3	**4**	**6**
First	B	D	D
Second	A	B	A
Third	C	A	C
Fourth	D	C	B

b. Election 2

Rank	Number of voters (13)		
	3	**4**	**6**
First	C	C	D
Second	B	B	A
Third	A	A	C
Fourth	D	D	B

c. Election 2

Rank	Number of voters (13)		
	3	**4**	**6**
First	B	B	D
Second	A	C	A
Third	C	A	C
Fourth	D	D	B

Word Search

Refer to page 384 of your text to obtain the Review Vocabulary. There are 9 hidden vocabulary words/expressions in the word search below. *Gibbard-Satterthwaite theorem* and *May's theorem for manipulability* were both omitted from the word search. It should be noted that spaces are removed as well as apostrophes.

```
S R M E Q E O Y C D R R R G O L E N F F B U X A U
I E B S A M M A E G S E P O C T T E E N J O I Q O
I N M L P W A S F S B X O D A R A P S R I A H C O
K I K S O T G J K S H S B I E T E B A J V D E I I
T B M F A H E A S O Z G V A D O F G V O L W D P X
V E L T E N N A M I O N P G P O H G E Q T V P E F
Q X Y H K P D I S I N G E N U O U S B A L L O T O
F F L A W E A K D O M I N A N C E T S A F B P I Y
U F P B G E M N C Q S E F M Z R O C P M N F H E S
I O S P R D A F O M X H O A X E Q E A E S L P B N
A F K U A C N D I E T A P L R R T S Z E W L R R T
P I O S S A I N E S C E Z C P E E O D D M N F E V
N T O F V I P O F B C T O I P K A O I E A S A A T
N O I T A L U P I N A M Q J A A E L L O C F N K L
M O H T O L L A B E R E C N I S N I V O S R R I D
S F K J A L A O H E G N A H C L A R E T A L I N U
H C U I C Y T I L I B A L U P I N A M P U O R G F
W K L E M G I V T Z A Y Q P H C I D L K H J A P N
I H T N T E O Z E X P E X N B T I S E O P W E O R
S E F Y A T N G N H O T I A G R R I W S E C S W A
E I S U V A O I E U O Z I C S I F I M B O O U E P
N G C S T R I S M L N P I Q G Z G I U F B H A R O
E T R J A T S E Q A E P O J I S F A L H Q I S E S
V H N E N S L E R F A G I M I T Y O T P E T R C I
M W F N O U F D Q I P P S O T P N Q R E F K R U S
```

1. _____

2. _____

3. _____

4. _____

5. _____

6. _____

7. _____

8. _____

9. _____

Chapter 11
Weighted Voting Systems

Chapter Objectives

Check off these skills when you feel that you have mastered them.

☐ Interpret the symbolic notation for a weighted voting system by identifying the quota, number of voters, and the number of votes each voter controls.

☐ Identify if a dictator exists in a given weighted voting system.

☐ Identify if a dummy exists in a given weighted voting system.

☐ Identify if a single voter has veto power in a given weighted voting system.

☐ Calculate the number of permutations of voters in a given weighted voting system.

☐ List the possible permutations for a three- or four-voter weighted voting system.

☐ Given a permutation of voters, identify the pivotal voter.

☐ Calculate the Shapley-Shubik index for a three- or four-voter weighted voting system.

☐ Identify winning coalitions by analyzing a given weighted voting system.

☐ Identify blocking coalitions by analyzing a given weighted voting system.

☐ When given a specific winning or blocking coalition from a weighted voting system, determine the critical voters.

☐ Determine the extra votes for a winning coalition.

☐ Calculate the Banzhaf power index for a given weighted voting system.

☐ Determine a specific value of C_k^n by using the combination formula as well as Pascal's triangle.

☐ Determine if two voting systems are equivalent and when given a voting system, find an equivalent system.

☐ Explain the difference between a winning coalition and a minimal winning coalition.

Guided Reading

Introduction

There are many settings, such as shareholder elections, in which people who are entitled to vote have varying numbers of votes. In such situations, the actual number of votes each can cast may not reflect the voter's *power*, that is, his ability to influence the outcome of the election. Several measures of power have been introduced and two of them are studied in this chapter, the *Banzhaf Power Index* and the *Shapley–Shubik Power Index*.

⌗ Key idea

A **weighted voting system** is one in which each voter has a number of votes, called his or her weight. The number of votes needed to pass a measure is called the quota. If the quota for a system with n voters is q, and the weights are $w_1, w_2, ..., w_n$, then we use the notation, $[q : w_1, w_2, ..., w_n]$.

ᔕ Example A

Consider the weighted voting system $[43 : 17, 28, 15, 22]$. Determine the number of voters and their weights. State the quota.

Solution

There are 4 voters. Their weights are 17, 28, 15, and 22. The quota is 43.

⌗ Key idea

A **dictator** is a voter whose weight is greater than or equal to the quota. Thus, if the dictator is in favor of a motion, it will pass. Moreover, a motion will fail if the dictator is against it, independent of how the other voters vote.

⌗ Key idea

A **dummy** is one whose vote will never be needed to pass or defeat any measure. Independent of how the dummy voter votes, the outcome will not change.

ᔕ Example B

List any dummy in the weighted voting system $[q : w_A, w_B, w_C, w_D] = [12 : 8, 6, 4, 1]$.

Solution

Voter D is a dummy voter. If D joined forces with any other single voter, their combined weight is not enough to win. If D joined forces with any other two voters, D's weight is not enough to alter the result. If a motion was to pass (A and B or A and C) then D has no influence, it will still pass. If the motion was to fail (B and C) then D has no influence, it will still fail. If D joins all three other voters, again D has no influence. The motion will pass.

⌗ Key idea

With different quotas, the distribution of power can be altered.

✑ Example C

Consider the weighted voting system $[q : w_A, w_B, w_C, w_D] = [13 : 8, 6, 4, 1]$.   Do we have any dummy voters?

Solution

There are no dummy voters.   In Example B, voter D was a dummy because he or she had no influence on the outcome.   With a quota of 13, voter D can make a motion pass by joining A and C.

🗝 Key idea

A single voter has **veto power** if no issue can pass without his or her vote.   Note that this is different from a dictator because the voter with veto power does not need to have a weight of the quota or greater.

✑ Example D

Consider the weighted voting system $[q : w_A, w_B, w_C, w_D] = [12 : 8, 6, 4, 1]$.   Does any voter have veto power.

Solution

Voter A has veto power.   Even if all the other voters B, C, and D all vote for a motion to pass, the sum of their weights is less than the quota.

Section 11.1   The Shapley-Shubik Power Index

🗝 Key idea

A **permutation** of voters is an ordering of all the voters.   There is $n!$ (n factorial) permutations of n voters where $n! = n \times (n-1) \times (n-2) \times ... \times 2 \times 1$.

✑ Example E

Consider the weighted voting system $[q : w_A, w_B, w_C, w_D, w_E] = [13 : 7, 5, 3, 2, 1]$.   How many permutation of voters are there?

Solution

There are 5 voters.   Thus, there are $5! = 5 \times 4 \times 3 \times 2 \times 1 = 120$ permutations of voters.

🗝 Key idea

The first voter in a permutation who, when joined by those coming before him or her, would have enough voting weight to win is called the **pivotal voter** of that permutation.   Each permutation has exactly one pivotal voter.

🗝 Key idea

The **Shapley–Shubik power index** of a voter is the fraction of the permutations in which that voter is pivotal.

⌇ Example F

Calculate the Shapley–Shubik power index for each of the voters in the weighted voting system $[q : w_A, w_B, w_C] = [8 : 6, 3, 2]$.

Solution

There are 3 voters. Thus, there are $3! = 3 \times 2 \times 1 = 6$ permutations of voters.

Permutations	Weights
A **B** C	6 **9** 11
A **C** B	6 **8** 11
B **A** C	3 **9** 11
B C **A**	3 5 **11**
C **A** B	2 **8** 11
C B **A**	2 5 **11**

Since A is the pivotal voter 4 times, B is pivotal 1 time, and C is pivotal 1 time, the Shapley–Shubik power index for this weighted system is $\left(\dfrac{4}{6}, \dfrac{1}{6}, \dfrac{1}{6} \right) = \left(\dfrac{2}{3}, \dfrac{1}{6}, \dfrac{1}{6} \right)$.

✏ Question 1

Calculate the Shapley–Shubik power index for each of the voters in the weighted voting system $[q : w_A, w_B, w_C, w_D] = [9 : 4, 3, 3, 1]$.

Answer

$\left(\dfrac{1}{3}, \dfrac{1}{3}, \dfrac{1}{3}, 0 \right)$

✏ Question 2

Calculate the Shapley–Shubik power index for each of the voters in the weighted voting system $[q : w_A, w_B, w_C, w_D] = [8 : 4, 3, 3, 1]$.

Answer

$\left(\dfrac{1}{2}, \dfrac{1}{6}, \dfrac{1}{6}, \dfrac{1}{6} \right)$

⌇ Key idea

When the number of voters (n) is large, then the number of permutations $(n!)$ is large. If there are 5 voters then there are $5! = 120$ permutations. If there are 6 voters then there are $6! = 720$ permutations and a direct calculation of the Shapley–Shubik index would be difficult. If, however, many of the voters have equal votes, it is possible to compute this index by counting the number of permutations.

⌘ Example G

Calculate the Shapley–Shubik power index for each of the voters in the weighted voting system $[q : w_A, w_B, w_C, w_D, w_E] = [4 : 3, 1, 1, 1, 1]$.

Solution

There are 5 voters. Thus, there are $5! = 5 \times 4 \times 3 \times 2 \times 1 = 120$ permutations of voters. A is pivotal in the following three types of permutations.

Permutations					Weights				
X_1	$\underline{A}$	X_2	X_3	X_4	1	$\underline{4}$	5	6	7
X_1	X_2	$\underline{A}$	X_3	X_4	1	2	$\underline{5}$	6	7
X_1	X_2	X_3	$\underline{A}$	X_4	1	2	3	$\underline{6}$	7

For each of these permutation types, there are $4! = 4 \times 3 \times 2 \times 1 = 24$ associated permutations. Thus, there is a total of $3 \times 24 = 72$ in which A is pivotal. Thus, the Shapley–Shubik power index for A is $\frac{72}{120} = \frac{3}{5}$. The remaining four voters share equally the remaining $1 - \frac{3}{5} = \frac{2}{5}$ of the power. Thus, each of them has an index $\frac{2}{5} \div 4 = \frac{2}{5} \times \frac{1}{4} = \frac{2}{20} = \frac{1}{10}$. The Shapley–Shubik power index for this weighted system is therefore $\left(\frac{3}{5}, \frac{1}{10}, \frac{1}{10}, \frac{1}{10}, \frac{1}{10} \right)$.

✎ Question 3

Calculate the Shapley–Shubik power index for each of the voters in the weighted voting system $[q : w_A, w_B, w_C, w_D] = [6 : 3, 2, 2, 2]$, without determining all possible permutations.

Answer

$\left(\frac{1}{4}, \frac{1}{4}, \frac{1}{4}, \frac{1}{4} \right)$

✎ Question 4

Calculate the Shapley–Shubik power index for each of the voters in the weighted voting system $[q : w_A, w_B, w_C, w_D, w_E, w_F] = [10 : 4, 4, 1, 1, 1, 1]$, without determining all possible permutations.

Answer

$\left(\frac{2}{5}, \frac{2}{5}, \frac{1}{20}, \frac{1}{20}, \frac{1}{20}, \frac{1}{20} \right)$

Section 11.2   The Banzhaf Power Index

☞ Key idea

A **coalition** is a set of voters that vote collectively in favor of or opposed to a motion. A **winning coalition** is a combination of voters with enough collective weight to pass a measure. A **blocking coalition** is a group of voters who have a sufficient number of votes to block a measure from passing. Note that a one-person blocking coalition is said to have veto power.

ᔕ Example H

List all of the winning coalitions in the weighted voting system given by $[q:w_A,w_B,w_C]=[7:5, 3, 2]$.

Solution

None of the voters have enough votes to individually make a motion pass.
The coalition $\{A, B\}$, with $5 + 3 = 8$ votes, exceeds the quota of 7.
The coalition $\{A, C\}$, with $5 + 2 = 7$ votes, matches the quota of 7.
The coalition $\{B, C\}$, with $3 + 2 = 5$ votes, is less than the quota of 7 and is thus a losing coalition.
The coalition $\{A, B, C\}$, with $5 + 3 + 2 = 10$ votes, exceeds the quota of 7.
Thus, the winning coalitions are $\{A, B\}$, $\{A, C\}$, and all three voters, $\{A, B, C\}$.

☞ Key idea

A voter is **critical** to a winning or blocking coalition if he can cause that coalition to lose by single-handedly changing his vote. Note that some winning coalitions have several critical voters, while others have none at all.

ᔕ Example I

In the weighted voting system $[q:w_A,w_B,w_C,w_D]=[12:8, 6, 4, 1]$, $\{A, B, D\}$ is a winning coalition. Find the critical voters in this coalition.

Solution

The coalition $\{A, B, D\}$, with $8 + 6 + 1 = 15$ votes, exceeds the quota of 12. If A drops out then the coalition $\{B, D\}$, with $6 + 1 = 7$ votes, has less than the quota of 12. If B drops out then the coalition $\{A, D\}$, with $8 + 1 = 9$ votes, has less than the quota of 12. If D drops out then the coalition $\{A, B\}$, with $8 + 6 = 14$ votes, has more than the quota of 12. Thus, A and B are critical voters.

☞ Key idea

A winning coalition must have a total weight of q or more. If w is the weight of the winning coalition, then that winning coalition has $w-q$ **extra votes**. Any voter that has a weight that exceeds the number of extra votes will be critical to that coalition.

ᔕ Example J

In the weighted voting system $[q:w_A,w_B,w_C,w_D]=[8:5, 5, 3, 2]$, coalition $\{B, C, D\}$ is a winning coalition. How many extra votes does it have, and which are the critical voters?

Solution

The coalition $\{B, C, D\}$, with $5 + 3 + 2 = 10$ votes, exceeds the quota of 8. Thus, there are 2 extra votes. Since the weights of B and C exceed 2, they are both are critical.

→ Key idea

If the combined weight of all voters is n voters, then a blocking coalition must have a weight more than $n - q$. Since votes are integer values, the blocking coalition must have a weight of at least $n - q + 1$.

→ Key idea

A voter's **Banzhaf power index** equals the number of distinct winning coalitions in which he is a critical voter. To determine this index for a voting system, perform the following.
- Make a list of the winning and blocking coalitions.
- Determine the number of extra votes a coalition has in order to identify the critical voters.

¤ Example K

Consider the weighted voting system $[q : w_A, w_B, w_C, w_D] = [12 : 8, 6, 4, 1]$. Find the Banzhaf power index for the system. Is there a dummy in this system?

Solution

The winning coalitions are those whose weights sum to 12 or more.

Winning coalition	Weight	Extra votes	Critical votes A	B	C	D
{A, B}	14	2	1	1	0	0
{A, C}	12	0	1	0	1	0
{A, B, C}	18	6	1	0	0	0
{A, B, D}	15	3	1	1	0	0
{A, C, D}	13	1	1	0	1	0
{A, B, C, D}	19	7	1	0	0	0
			6	2	2	0

The combined weight of all voters is 19. A blocking coalition must have a weight of $19 - 12 + 1 = 8$ or more. $[q : w_A, w_B, w_C, w_D] = [12 : 8, 6, 4, 1]$.

Blocking coalition	Weight	Extra votes	Critical votes A	B	C	D
{A}	8	0	1	0	0	0
{A, B}	14	6	1	0	0	0
{A, C}	12	4	1	0	0	0
{A, D}	9	1	1	0	0	0
{B, C}	10	2	0	1	1	0
{A, B, C}	18	10	0	0	0	0
{A, B, D}	15	7	1	0	0	0
{A, C, D}	13	5	1	0	0	0
{B, C, D}	11	3	0	1	1	0
{A, B, C, D}	19	11	0	0	0	0
			6	2	2	0

Adding the number of times each voter is critical in either a winning coalition or blocking coalition, the Banzhaf index of this system is $(12, 4, 4, 0)$. D has an index of 0 and is a dummy.

⌐ Key idea

The number of winning coalitions in which a voter is critical is equal to the number of blocking coalitions in which the same voter is critical. This is known as **winning/blocking duality**. Thus, to calculate a voter's Banzhaf power index, one can double the number of times that voter is critical in a winning coalition.

✎ Question 5

Consider the weighted voting system $[q : w_A, w_B, w_C, w_D] = [11 : 7, 5, 3, 2]$. Find the Banzhaf power index for the system. Is there a dummy in this system?

Answer

$(10, 6, 2, 2)$; No.

⌐ Key idea

A **voting combination** is a record of how the voters cast their votes for or against a given proposition. Since there are only two outcomes, in favor or against, n voters can have 2^n voting combinations. If a vote against (no) a proposition is cast, that voter can be represented by a 0. If a vote for (yes) a proposition is cast, that voter can be represented by a 1. Thus, a voting combination can be sequence of 0's and/or 1's.

⌐ Key idea

A sequence made up of 0's and/or 1's (**bits**) can represent a **binary number** or a base-2 number. One should be able to express a number (base 10) in binary form (base 2) and vice versa. Generating powers of 2 is often helpful in such conversions.

n	0	1	2	3	4	5	6	7
2^n	1	2	4	8	16	32	64	128

n	8	9	10	11	12	13	14
2^n	256	512	1024	2048	4096	8192	16,384

↬ Example L

a) Express the binary number 1001110 in a standard form (base 10).

b) Express 10,134 in binary notation.

Solution

a) $1001110 = 2^6 + 2^3 + 2^2 + 2^1 = 64 + 8 + 4 + 2 = 78$.

b) Since 2^{13} represents the largest power of 2 that doesn't exceed 10134, we start there.

$$10,134 - 8192 = 10,134 - 2^{13} = 1942$$
$$1942 - 1024 = 1942 - 2^{10} = 918$$
$$918 - 512 = 918 - 2^9 = 406$$
$$406 - 256 = 406 - 2^8 = 150$$
$$150 - 128 = 150 - 2^7 = 22$$
$$22 - 16 = 22 - 2^4 = 6$$
$$6 - 4 = 6 - 2^2 = 2$$
$$2 - 2 = 2 - 2^1 = 0$$

Thus, the nonzero bits are $b_{13}, b_{10}, b_9, b_8, b_7, b_4, b_2$, and b_1. Thus, we have the following.

$$(10,134)_2 = 10011110010110$$

🔑 Key idea

The number of voting combinations with n voters and exactly k "yes" votes can be determined by finding C_k^n. This value can be found by using the **combination formula**, $C_k^n = \dfrac{n!}{k!(n-k)!}$. Since there is only one way to obtain all "no" votes or all "yes" votes, it should be noted that $C_0^n = C_n^n = 1$. This does agree with the combination formula.

$$C_0^n = \frac{n!}{0!(n-0)!} = \frac{n!}{1 \cdot n!} = 1 \text{ and } C_n^n = \frac{n!}{n!(n-n)!} = \frac{n!}{n!\,0!} = \frac{n!}{n! \cdot 1} = 1$$

⌇ Example M

Use the combination formula to find C_5^7.

Solution

Since $C_k^n = \dfrac{n!}{k!(n-k)!}$, we have $C_5^7 = \dfrac{7!}{5!(7-5)!} = \dfrac{7!}{5!\,2!} = \dfrac{7 \times 6}{2 \times 1} = \dfrac{42}{2} = 21.$

🔑 Key idea

The **duality formula for combinations** is $C_k^n = C_{n-k}^n$, and the **addition formula** is $C_{k-1}^n + C_k^n = C_k^{n+1}$. Using the addition formula, the pattern known as **Pascal's Triangle** can be justified. The first 11 rows are displayed below.

```
                        1
                      1   1
                    1   2   1
                  1   3   3   1
                1   4   6   4   1
              1  5  10   10   5  1
            1  6  15  20  15   6  1
          1  7  21  35  35  21  7  1
        1  8  28  56  70  56  28  8  1
      1  9  36  84  126 126 84  36  9  1
   1  10  45 120 210 252 210 120 45 10  1
```

Pascal's triangle is useful in finding values of C_k^n, where n is relatively small.

⌇ Example N

Use Pascal's triangle to find C_5^7.

Solution

With the very top 1 being the starting place (0th row), go down to the 7th row and to the 5th entry (starting with the 0th entry).

```
               1
             1   1
           1   2   1
         1   3   3   1
       1   4   6   4   1
     1  5  10   10   5  1
   1  6  15  20  15   6  1
 1  7  21  35  35  (21)  7  1
```

Thus, $C_5^7 = 21$.

⌇ Key idea

When there are many voters, the number of winning coalitions can be very large, and calculating the Banzhaf index will then be cumbersome. However, there are settings in which most, but not all, of the voters have equal weights. In such situations, we can compute the Banzhaf index by means of combinations, using the numbers C_k^n.

⌇ Example 0

Consider the weighted voting system $[q:w_A,w_B,w_C,w_D,w_E]=[4:2,1,1,1,1]$. Find the Banzhaf power index for the system.

Solution

A is a critical voter in just two types of winning coalitions. One in which there are no extra votes or 1 extra vote. X_i indicates a weight-one voter, where $i=1, 2, 3$, or 4.

Winning coalition	Weight	Extra votes
$\{A,X_1,X_2\}$	4	0
$\{A,X_1,X_2,X_3\}$	5	1

In the first situation, the two additional voters are drawn from the four other voters, and there are $C_2^4=6$ ways of choosing these two voters. Similarly, in the second case, there are $C_3^4=4$ ways of choosing three voters from among four. Hence, A is critical to 10 winning coalitions, and there are 10 blocking coalitions in which he is also critical, so that his Banzhaf index is 20. Now let us consider one of the voters with just one vote, say B, who is also critical in two types of winning coalitions.

Winning coalition	Weight	Extra votes
$\{A,B,X_1\}$	4	0
$\{B,X_1,X_2,X_3\}$	4	0

In the first situation, the additional voter is drawn from three other voters. and there are $C_1^3=3$ ways of choosing this voter. Similarly, in the second case, there is $C_3^3=1$ way of choosing three voters from among three. This, together with an equal number of blocking coalitions in which he is critical, yields a Banzhaf index of 8.

Thus, the Banzhaf power index for the system $(20,8,8,8,8)$.

✎ Question 6

Consider the weighted voting system $[q:w_A,w_B,w_C,w_D,w_E,w_F]=[5:2,2,1,1,1,1]$. Find the Banzhaf power index for the system.

Answer

$(30,30,14,14,14,14)$

Section 11.3   Comparing Voting Systems

⊶ Key idea

Many voting systems are not presented as weighted voting systems, but are equivalent to weighted systems. Two voting systems are **equivalent** if there is a way to exchange all voters from the first system with voters of the second while maintaining the same winning coalitions.

⊶ Key idea

A **minimal winning coalition** is one in which each voter is critical to the passage of a measure; that is, if anyone defects, then the coalition is turned into a losing one.

⌢ Example P

Consider the weighted voting system $[q : w_A, w_B, w_C] = [7 : 5, 3, 2]$.   Find the minimal winning coalitions.

Solution

The winning coalitions are {A, B}, {A, C}, and {A, B, C} with weights 8, 7, and 10, respectively. Of these only {A, B} and {A, C} are minimal.

⊶ Key idea

A voting system can be described completely by stating its minimal winning coalitions. There are three requirements of this list of winning coalitions.

- You must have at least one coalition on the list; otherwise, a motion can't pass.

- A minimal coalition cannot be contained in another minimal one. This would contradict the idea that this is a <u>minimal</u> winning coalition.

- Every pair of minimal winning coalitions should overlap; otherwise, two opposing motions can pass.

⌢ Example Q

A small club has 5 members. A is the President, B is the Vice President, and C, D, and E are the ordinary members. The minimal winning coalitions are A and B, A and any two of the ordinary members, and B and all three of the ordinary members. Express this situation as an equivalent weighted voting system.

Solution

There are many answers possible. Consider assigning a weight of 1 to each regular committee member. Then assign a weight to the Vice President and finally to the President with both based on the voting requirements. One possible answer is therefore $[q : w_A, w_B, w_C, w_D, w_E] = [5 : 3, 2, 1, 1, 1]$. Another possibility would be to assign a weight of 2 to each regular committee member. We could therefore have the following $[q : w_A, w_B, w_C, w_D, w_E] = [9 : 5, 4, 2, 2, 2]$.

Homework Help

Exercises 1 – 4
Carefully read the Introduction before responding to these exercises. Be familiar with how to read a weighted voting system and know the key words such as dictator, veto power, and dummy.

Exercises 5 – 6
Carefully read Section 11.1 before responding to these exercises. Be familiar with the role of a pivotal voter.

Exercises 7 – 9
Carefully read Section 11.1 before responding to these exercises. When determining permutations in which a voter is pivotal, first determine under what conditions (like position) a voter is pivotal. Then systematically list those permutations out in Exercise 7. When calculating the Shapley-Shubik power index of this weighted voting system, recall there are $n!$ permutations, where n is the number of voters in the system.

Exercise 10
Calculate the ratio of Bush to Kerry votes if 1000 were taken from Bush and 1000 were added to Kerry. Interpret the results given in Table 11.2 in Section 11.1.

Exercise 11
Be consistent with the labeling of the voters. One way is to let the first bit on the left correlate to Voter A.

Exercise 12
Carefully read Section 11.2 before responding to this exercise. There are eight winning coalitions. You may find the following table helpful. Place a 1 in the voter's column if he or she is critical to the winning coalition. Otherwise, place a 0.

Winning coalition	Weight	Extra votes	Critical votes			
			A	B	C	D

Determine the weight needed for a coalition to be a blocking coalition. There are six blocking coalitions in which the weight-30 voter (Voter A) is critical. To determine a blocking coalition's dual winning coalition in which the same voter is critical, determine the voters that don't appear in that blocking coalition and then include the weight-30 voter (Voter A).

Exercise 13

Carefully read Section 11.2 before responding to this exercise. Let's call the voters A, B, C, and D. This weighted voting system can be written as $[q : w_A, w_B, w_C, w_D] = [q : 30, 25, 24, 21]$.

(a) $[q : w_A, w_B, w_C, w_D] = [52 : 30, 25, 24, 21]$ Copy the table of coalitions we made for Exercise 12, reducing the extra votes of each by 1. label any losing coalition.

Winning coalition	Weight	Extra votes	Critical votes			
			A	B	C	D

Double to account for blocking coalitions.

(b) $[q : w_A, w_B, w_C, w_D] = [55 : 30, 25, 24, 21]$

We copy the table from Part (a), dropping the losing coalition and reducing quotas by 3. One more coalition will lose.

Winning coalition	Weight	Extra votes	Critical votes			
			A	B	C	D

Double to account for blocking coalitions.

(c) $[q : w_A, w_B, w_C, w_D] = [58 : 30, 25, 24, 21]$

We copy the table from Part (b), dropping the losing coalition and reducing quotas by 3. One more coalition will lose.

Winning coalition	Weight	Extra votes	Critical votes			
			A	B	C	D

Double to account for blocking coalitions.

Continued on next page

Exercise 13 continued

(d) $[q:w_A,w_B,w_C,w_D]=[73:30,\,25,\,24,\,21]$

We copy the table from Part (c), dropping the losing coalition and reducing quotas by 15. One more coalition will lose. One of the voters will acquire veto power.

Winning coalition	Weight	Extra votes	Critical votes			
			A	B	C	D

Double to account for blocking coalitions.

(e) $[q:w_A,w_B,w_C,w_D]=[76:30,\,25,\,24,\,21]$

We copy the table from Part (d), dropping the losing coalition and reducing quotas by 3. One more coalition will lose.

Winning coalition	Weight	Extra votes	Critical votes			
			A	B	C	D

Double to account for blocking coalitions.

(f) $[q:w_A,w_B,w_C,w_D]=[79:30,\,25,\,24,\,21]$

We copy the table from Part (e), dropping the losing coalition and reducing quotas by 3. One more coalition will lose. In this system, one f the voters is a dummy.

Winning coalition	Weight	Extra votes	Critical votes			
			A	B	C	D

Double to account for blocking coalitions.

(g) $[q:w_A,w_B,w_C,w_D]=[82:30,\,25,\,24,\,21]$

Only one winning coalition is left, with 18 extra votes. This is less than the weight of each participant. All voters are critical. In this system, a unanimous vote is required to pass a motion.

Winning coalition	Weight	Extra votes	Critical votes			
			A	B	C	D

Double to account for blocking coalitions.

Exercise 14
Carefully read Section 11.2 before responding to this exercise. In each system, the voters will be denoted A, B, ...

In Parts (a) – (d), determine the number of extra votes for each winning coalitions and how many times a voter is critical. Don't forget to double the number in order to determine the Banzhaf power index. In Part (e), there are eight winning coalitions. The following table may be helpful.

Winning coalition	Weight	Extra votes	Critical votes			
			A	B	C	D

Exercise 15
Carefully read Section 11.2 before responding to this exercise. A table of powers of 2 may be helpful. See Page 272 of this *Guide*.

Exercises 16 – 17
Carefully read Section 11.2 before responding to these exercises. You will need to use the formula $C_k^n = \dfrac{n!}{k!(n-k)!}$, where it is defined. Try to simplify as much as possible by canceling. Also, you may be able to use the duality formula.

Exercise 18
Carefully read Section 11.2 before responding to this exercise. Example 11.16 may be helpful. In Part (f), convert Table 11.2 into a table of percentages. The following blank table may be helpful.

Supervisor from	Population	Number of votes	Banzhaf power index	
Quota			65	72
Hempstead (Presiding) Hempstead				
North Hempstead				
Oyster Bay				
Glen Cove				
Long Beach				

Exercise 19 – 26
Carefully read Sections 11.2 and 11.3 before responding to these exercises

Exercise 27
Carefully read Section 11.3 before responding to this exercise. All four-voter systems can be presented as weighted voting systems. As indicated in the exercise, there are a total of nine. An example of one of the nine is the weighted voting system $[q:w_A,w_B,w_C,w_D]=[4:3,1,1,1]$ with minimal winning coalitions of $\{A, B\}$, $\{A, C\}$, and $\{A, D\}$.

Exercise 28
Carefully read Section 11.2 before responding to this exercise. Call the voters A, B, C, and D. This weighted voting system can be written as $[q:w_A,w_B,w_C,w_D]=[51:48, 23, 22, 7]$. The following table may be helpful.

Winning coalition	Weight	Extra votes	Losing coalition	Weight	Votes needed

Exercise 29
Carefully read Section 11.3 before responding to this exercise. Call the voters A, B, C, and D and determine the minimal winning coalitions. Compare with each part.

Exercise 30
Carefully read Section 11.2 before responding to this exercise. To determine the Banzhaf index, refer to the table in answer 28(a). To determine the Shapley-Shubik power index, consider the permutations in which D is pivotal.

Exercise 31
Carefully read Sections 11.2 and 11.3 before responding to this exercise. In Part (b) use combinations to determine the winning committees in which the chairperson is critical. In Part (c) divide the permutations into 9 groups, according to the location of the chairperson.

Exercise 32
Carefully read Sections 11.1 - 11.3 before responding to this exercise. In Part (a) there are three separate groupings that will yield a minimal winning coalition. In Part (b) use combinations to determine the number of ways each type of person is critical in a winning coalition.

Exercise 33
Carefully read Sections 11.1 - 11.3 before responding to this exercise. There are three separate groupings that will yield a minimal winning coalition.

Exercise 34
Carefully read Sections 11.1 - 11.3 before responding to this exercise. In Part (a) determine the conditions in which one would have a minimal winning coalition. In Part (b), first determine the Banzhaf index of each permanent member. Use combinations to find this number. In Part (c) it will end up that each permanent member is 63 times as powerful as each non-permanent member.

Exercise 35 – 36
Carefully read Sections 11.1 - 11.3 before responding to these exercises.

Exercise 37
Carefully read the Section 11.1 before responding to this exercise

Exercise 38
Carefully read the Section 11.3 before responding to this exercise

Try to assign weights to make this a weighted voting system. Give each recent graduate (RG) a weight of 1, and let X be the weight of each of the rich alumni (RA). The quota will be denoted Q. Determine the minimal winning coalitions.

Exercise 39
Let r_1, r_2, and r_M denote the rations for the two districts and the state as a whole, respectively, and let y_1, y_2, and y_M be the numbers of votes cast for the Kerry-Edwards ticket in each entity.

Do You Know the Terms?

Cut out the following 28 flashcards to test yourself on Review Vocabulary. You can also find these flashcards at http://www.whfreeman.com/fapp7e.

Chapter 11 **Weighted Voting Systems** **Addition formula**	**Chapter 11** **Weighted Voting Systems** **Banzhaf power index**
Chapter 11 **Weighted Voting Systems** **Bit**	**Chapter 11** **Weighted Voting Systems** **Binary number**
Chapter 11 **Weighted Voting Systems** **Blocking coalition**	**Chapter 11** **Weighted Voting Systems** C^n_k
Chapter 11 **Weighted Voting Systems** **Coalition**	**Chapter 11** **Weighted Voting Systems** **Critical voter**
Chapter 11 **Weighted Voting Systems** **Dictator**	**Chapter 11** **Weighted Voting Systems** **Duality formula**

A count of the winning or blocking coalitions in which a voter is a critical member. This is a measure of the actual voting power of that voter.	$$C_k^{n+1} = C_{k-1}^n + C_k^n$$
The expression of a number in base-2 notation	A binary digit: 0 or 1.
The number of voting combinations in a voting system with n voters, in which k voters say "Yes" and $n-k$ voters say "No." This number, referred to as "n-choose-k," is given by the formula $C_k^n = \dfrac{n!}{k! \times (n-k)!}$.	A coalition in opposition to a measure that can prevent the measure from passing.
A member of a winning coalition whose vote is essential for the coalition to win, or a member of a blocking coalition whose vote is essential for the coalition to block.	The set of participants in a voting system who favor, or who oppose a given motion. A coalition may be empty (if, for example, the voting body unanimously favors a motion, the opposition coalition is empty); it may contain some but not all voters, or it may consist of all the voters.
$$C_{n-k}^n = C_k^n$$	A participant in a voting system who can pass any issue even if all other voters oppose it and block any issue even if all other voters approve it.

Chapter 11
Weighted Voting Systems

Dummy

Chapter 11
Weighted Voting Systems

Equivalent voting systems

Chapter 11
Weighted Voting Systems

Extra votes

Chapter 11
Weighted Voting Systems

Extra-votes principle

Chapter 11
Weighted Voting Systems

Factorial

Chapter 11
Weighted Voting Systems

Losing coalition

Chapter 11
Weighted Voting Systems

Minimal winning coalition

Chapter 11
Weighted Voting Systems

Pascal's triangle

Chapter 11
Weighted Voting Systems

Permutation

Chapter 11
Weighted Voting Systems

Pivotal voter

Two voting systems are equivalent if there is a way for all the voters of the first system to exchange places with the voters of the second system and preserve all winning coalitions.

A participant who has no power in a voting system. A dummy is never a critical voter in any winning or blocking coalition and is never the pivotal voter in any permutation.

The critical voters in the coalition are those whose weights are more than the extra votes of the coalition. For example, if a coalition has 12 votes and the quota is 9, there are 3 extra votes. The critical voters in the coalition are those with more than 3 votes.

The number of votes that a winning coalition has in excess of the quota.

A coalition that does not have the voting power to get its way.

The number of permutations of n voters (or n distinct objects) which is symbolized $n!$.

A triangular pattern of integers, in which each entry on the left and right edges is 1, and each interior entry is equal to the sum of the two entries above it. The entry that is located k units from the left edge, on the row n units below the vertex, is C_k^n.

A winning coalition that will become losing if any member defects. Each member is a critical voter.

The first voter in a permutation who, with his or her predecessors in the permutation, will form a winning coalition. Each permutation has one and only one pivotal voter.

A specific ordering from first to last of the elements of a set; for example, an ordering of the participants in a voting system.

Chapter 11
Weighted Voting Systems

Power Index

Chapter 11
Weighted Voting Systems

Quota

Chapter 11
Weighted Voting Systems

Shapley –Shubik power index

Chapter 11
Weighted Voting Systems

Veto power

Chapter 11
Weighted Voting Systems

Voting Combination

Chapter 11
Weighted Voting Systems

Weight

Chapter 11
Weighted Voting Systems

Weighted voting system

Chapter 11
Weighted Voting Systems

Winning coalition

The minimum number of votes necessary to pass a measure in a weighted voting system.

A numerical measure of an individual voter's ability to influence a decision, the individual's voting power.

A voter has veto power if no issue can pass without his or her vote. A voter with veto power is a one-person blocking coalition.

A numerical measure of power for participants in a voting system. A participant's Shapley – Shubik index is the number of permutations of the voters in which he or she is the pivotal voter, divided by the number of permutations ($n!$ if there are n participants).

The number of votes assigned to a voter in a weighted voting system, or the total number of votes of all voters in a coalition.

A list of voters indicating the vote on an issue. There are a total of 2^n combinations in an n-element set, and C_k^n combinations with k "yes" votes and $n - k$ "no" votes.

A set of participants in a voting system who can pass a measure by voting for it.

A voting system in which each participant is assigned a voting weight (different participants may have different voting weights). A quota is specified, and if the sum of the voting weights of the voters supporting a motion is at least equal to that quota, the motion is approved.

Practice Quiz

1. What would be the quota for a voting system that has a total of 30 votes and uses a simple majority quota?
   a. 15
   b. 16
   c. 30

2. A small company has three stockholders: the president and vice president hold 6 shares each, and a long-time employee holds 2 shares. The company uses a simple majority voting system. Which statement is true?
   a. The long time employee is a dummy voter.
   b. The employee is not a dummy, but has less power than the officers.
   c. All three shareholders have equal power.

3. Which voters in the weighted voting system [12: 11, 5, 4, 2] have veto power?
   a. no one
   b. A only
   c. Both A and B

4. The weighted voting system [12: 11, 5, 4, 2] has how many winning coalitions?
   a. 1
   b. 7
   c. 8

5. If there are three voters in a weighted voting system, how many distinct coalitions of voters can be formed?
   a. 6
   b. 8
   c. 9

6. Given the weighted voting system [6: 4, 3, 2, 1], find the number of extra votes of the coalition $\{A, B, C\}$.
   a. 1
   b. 2
   c. 3

7. For the weighted voting system [10: 4, 4, 3, 2], which of the following is true?
   a. A has more power than C.
   b. A and C have equal power.
   c. C has more power than D.

8. What is the value of C_4^7?
   a. 210
   b. 35
   c. 28

9. Find the Banzhaf power index for voter B in the weighted voting system [12: 11, 5, 4, 2].
    a. 4
    b. 8
    c. 10

10. Calculate the Shapley-Shubik power index for voter B in the system [12: 11, 5, 4, 2].
    a. $\dfrac{1}{12}$
    b. $\dfrac{3}{12}$
    c. $\dfrac{8}{12}$

Word Search

Refer to pages 421 – 422 of your text to obtain the Review Vocabulary. There are 26 hidden vocabulary words/expressions in the word search. *Coalition, blocking coalition, weighted voting system*, and *weight* all appear separately. C_k^n and *Voting Combination* do not appear. Spaces and apostrophes are removed.

```
N O I T A T U M R E P F F T A T A R G I X N P G A
R E W O P O T E V Z S A R I E J L E B D C O O N E
E L G N A I R T S L A C S A P V F X X I E I E A W
Y E N C I V N S W O S T O D I E Z E E C C T S D X
E X E N I V N Y P V E O D A V T G D E T L I K D S
V T L O E B O S S D T R A T O U Q N K A E L I I R
H S I I L I I G G M O I N B T K M I F T P A T T G
C B Z T P N T N X R V A V L A A L R C O O O V I S
R A E I I A I I C B A L T O L J S E U R B C S O B
I A N L C R L T O I R S N C V H R W L P A G S N A
T D T A N Y A O M A T Z E K O E J O L F N N L F T
I D P O I N O V V L X T H I T M H P W W Z I R O W
C F A C R U C D I U E P H N E A H K E G H N F R R
A T K G P M G E N M U B R G R E P I N C A N E M E
L I E N S B N T A R T F G C I O G B R O F I S U J
V O P I E E I H T O R M O O W H N U T A P W E L R
O N H N T R S G I F G K O A T E W H V L O L B A L
T Y C N O X O I O Y Y E N L S S I S E I W A A O E
E Q U I V A L E N T V O T I N G S Y S T E M S V I
R K A W A Q C W D I N L S T J F Y E H I R I I S N
W M E J R B S V A L O S P I T C T L E O I N D K L
C U R Z T N Y D D A E B M O I C A P R N N I A D W
T L H E X Z I P D U M M Y N L B P A N E D M F S P
E Q V T E E P C P D N A T D R T E H Q S E X E Y G
A D R N C X E D N I R E W 0 P W J S O H X F R E E
```

1. _____   10. _____   19. _____

2. _____   11. _____   20. _____

3. _____   12. _____   21. _____

4. _____   13. _____   22. _____

5. _____   14. _____   23. _____

6. _____   15. _____   24. _____

7. _____   16. _____   25. _____

8. _____   17. _____   26. _____

9. _____   18. _____

Chapter 12
Electing the President

Chapter Objectives

Check off these skills when you feel that you have mastered them.

- [ ] Define a voter distribution.

- [ ] Describe a symmetric voter distribution.

- [ ] Identify the modes in a voter distribution.

- [ ] Describe a unimodal voter distribution.

- [ ] Know how to find the median and mean of a voter distribution.

- [ ] Explain the meaning of a candidate's maximin position.

- [ ] Explain the meaning of an equilibrium position.

- [ ] State the median-voter theorem.

- [ ] Understand the meaning of the 1/3-separation obstacle.

- [ ] Understand the meaning of the 2/3-separation opportunity.

- [ ] Describe the bandwagon effect.

- [ ] Describe the spoiler problem.

- [ ] Compare sincere voting to strategic voting.

- [ ] State the poll assumption.

- [ ] Determine the Condorcet winner of an election, if one exists.

- [ ] Explain how dichotomous preferences affect a voter's dominant strategy.

- [ ] Find the expected popular vote (*EPV*) and the expected electoral vote (*EEV*).

Guided Reading

Introduction

Presidential elections can be modeled mathematically as games, in which rules and optimal strategies can change from phase to phase. Mathematical models can be used to determine good campaign strategies at each phase, and to predict the effects that election reforms might have on campaign strategies.

Section 12.1   Spatial Models for Two-Candidate Elections

☞ Key idea

Voters respond to the positions that candidates take on election issues. Attitudes of voters are represented along a left-right continuum, where the left indicates a very liberal outlook and the right represents a very conservative outlook. A **voter distribution** is a curve that shows the number of voters who have attitudes at each point along the horizontal axis. The height of the curve indicates how many voters have attitudes at that point.

☞ Key idea

The **mode** of a voter distribution is a peak in the curve. A distribution with only one such peak is **unimodal**.

☞ Key idea

The **median** M is the point on the horizontal axis where half of the voters have attitudes to the left of the point M and half of the voters have attitudes to the right of the point M. If the curve to the left of M is a mirror image of the curve to the right of M, the distribution is **symmetric**.

☞ Key idea

A candidate's position is **maximin** if there is no other position that will guarantee more votes for that candidate, no matter what position the other candidate may take.

☞ Key idea

Once both candidates have chosen their positions, the positions are in **equilibrium** if neither candidate has motivation to change his or position independently.

☞ Key idea

The **median-voter theorem** states that for a two-candidate election with an odd number of voters, M is the unique equilibrium position.

☞ Key idea

In an election with n voters taking k different positions along the horizontal axis, we can call each position i and indicate the number of voters taking position i with the notation n_i. Using l_i as the location of position i along the horizontal axis, we can find the mean $\bar{l}$ of a voting distribution using

$$\bar{l} = \frac{1}{n}\sum_{i=1}^{k} n_i l_i.$$

᠗ Example A

Consider the following distribution of 21 voters at 5 different positions over the interval [0, 1].

Position i	1	2	3	4	5
Location (l_i) of position i	0.1	0.3	0.5	0.7	0.8
Number of voters (n_i) at position i	1	5	5	6	4

What is the mean position?

Solution

Using the formula, we have the following.

$$\bar{l} = \frac{1}{21} \sum_{i=1}^{5} n_i l_i$$

$$\bar{l} = \frac{1}{21}\left[1(0.1) + 5(0.3) + 5(0.5) + 6(0.7) + 4(0.8)\right] = \frac{1}{21}\left[0.1 + 1.5 + 2.5 + 4.2 + 3.2\right] = \frac{1}{21}(11.5) \approx 0.55$$

✎ Question 1

Consider the following distribution of 16 voters at 4 different positions over the interval [0, 1].

Position i	1	2	3	4
Location (l_i) of position i	0.1	0.3	0.5	0.7
Number of voters (n_i) at position i	2	5	5	4

What is the mean position? Round to the nearest hundredth.

Answer

0.44

Section 12.2   Spatial Models for Multi-Candidate Elections

᠊ Key idea

Suppose a distribution of voters is symmetric and unimodal, and the first two candidates have chosen different positions A and B that are equidistant from the median such that A is below the median, B is above the median, and no more than 1/3 of the voters lie between A and B. Then a third candidate cannot possibly take a position C that will displace A and B and allow C to win. **This is the 1/3-separation obstacle.**

᠊ Key idea

Suppose a distribution of voters is symmetric and unimodal, and the first two candidates have chosen different positions A and B that are equidistant from the median such that A is below the median, B is above the median, and 2/3 of the voters lie between A and B. Then a third candidate can win the election by taking a position C that lies at M, halfway between A and B. **This is the 2/3-separation opportunity.**

᠊ Key idea

Suppose a distribution of voters is uniform over $[0,1]$. Candidates A and B have already entered the election and anticipate a third candidate C to enter the election later. Candidates A and B should choose positions at 1/4 and at 3/4. Then there is no position at which a third candidate C can enter such that C can win, so these positions are the optimal entry positions for two candidates anticipating a third entrant.

Section 12.3   Winnowing the Field

☞ **Key idea**

The **bandwagon effect** provokes voters to vote for the presumed winner of an election, independent of that candidate's merit.

☞ **Key idea**

The **spoiler problem** is caused by a candidate that cannot win but influences the election in such a way that an otherwise winning candidate does not win.

Section 12.4   What Drives Candidates Out?

☞ **Key idea**

Polls publicly indicate the standings of candidates in an election. This can affect the outcome of the election.

☞ **Key idea**

In plurality voting, each voter votes for only one candidate and the candidate with the most votes wins the election. A voter is voting **sincerely** if the voter casts a vote for his or her favorite candidate.

☞ **Key idea**

The poll assumption presumes that voters will change their sincere voting strategy to select one of the top two candidates as indicated by the poll, voting for the one they prefer.

☞ **Key idea**

A candidate that can defeat each of the other candidates in a pairwise contest is called a Condorcet winner.

↷ **Example B**

Assume there are three classes of voters that rank three candidates as follows:

I.	3:	A	B	C
II.	12:	C	A	B
III.	10:	B	A	C

Which candidate is the Condorcet winner?

Solution

There is a total of 25 voters. In a pairwise comparison of A to B, A is preferred by the 3 voters in class I and the 12 voters in class II for a total of 15 votes. Only 10 people prefer B to A, so A beats B in a pairwise contest. Comparing A to C, we see that 3 voters in class I and 10 voters in class III prefer A to C, for a total of 13 votes. Only 12 people prefer C to A, so A also beats C in a pairwise comparison. Because A beats both B and C, there is no need to compare B to C. Candidate A is the Condorcet winner.

☞ **Key idea**

A Condorcet winner will always lose if he or she is not one of the top two candidates as indicated by the poll, based on the poll assumption. A Condorcet winner will always win if he or she *is* one of the top two candidates as indicated by the poll.

Section 12.5   Election Reform: Approval Voting

☞ **Key idea**

Approval voting allows voters to cast one vote each for as many candidates as they wish. The candidate with the most approval votes wins the election.

☞ **Key idea**

A voter may vote for a candidate that is not his or her first choice in an attempt to elect an acceptable candidate if his or her first choice is not likely to win. This is called **strategic voting**.

☞ **Key idea**

If approval voting is being used in a three-candidate election, it is never rational for a voter to vote only for a second choice. He or she should vote for a first choice as well.

☞ **Key idea**

A voter with **dichotomous preferences** divides a set of candidates into a preferred subset and a nonpreferred subset, and is indifferent among all candidates in each subset. In this case the voter can make a rational choice by choosing a strategy that is at least as good as, if not better than, any other strategy for that voter. This strategy is called a **dominant strategy**.

☞ **Key idea**

Under approval voting, a Condorcet winner will always win if all voters have dichotomous preferences and choose their dominant strategies.

◠ **Example C**

Suppose there are four classes of voters that rank four candidates as shown below. For each class of voters, the preferred subset of candidates is enclosed in the first set of parentheses and the non-preferred subset in the second set of parentheses. Thus, the 4 class I voters prefer A and C, between whom they are indifferent, to B and D, between whom they are also indifferent:

$$
\begin{array}{llllll}
\text{I.} & 4: & (A & C) & (B & D) \\
\text{II.} & 3: & (A & B & D) & (C) \\
\text{III.} & 7: & (B) & (A & C & D) \\
\text{IV.} & 5: & (C) & (A & B & D)
\end{array}
$$

Assuming that each class of voters chooses its dominant strategy, who wins?

Solution

Candidate A will get 4 votes from the first class of voters and 3 votes from the second class of voters, for a total of 7 votes. Candidate B will get 3 votes from the second class of voters and 7 votes from the third class of voters, for a total of 10 votes. Candidate C will get 4 votes from the first class of voters and 5 votes from the fourth class of voters, for a total of 9 votes. Candidate D gets only the 3 votes from the second class of voters. Candidate B has the most votes, so B wins.

Section 12.6   The Electoral College

☞ **Key idea**

The **Electoral College** provides 538 electoral votes, and a candidate needs 270 of these votes to win. Each state gets one vote for each of its two senators and one vote for each of its representatives in the House of Representatives. The District of Columbia also gets 3 electoral votes.

☞ Key idea

A state in which an election outcome is expected to be close is a toss-up state. The **expected popular vote (*EPV*)** of the Democratic candidate in toss-up states, EPV_D, is given by

$$EPV_D = \sum_{i=1}^{t} n_i p_i.$$

where t is the number of toss-up states, n_i is the number of voters in toss-up state i, and p_i is the probability that a voter in toss-up state i votes for the Democrat candidate.

☞ Key idea

The Democratic candidate should allocate his or her resources in proportion to the size of each state. If $N = \sum_{i=1}^{t} n_i$ is the total number of voters in the toss-up states and $D = \sum_{i=1}^{t} d_i$ is the sum of the Democrat's expenditures across the toss-up states, then the Democrat should allocate $d_i* = (n_i / N)D$ to each toss-up state i. This assumes that the Republican behaves similarly by following a strategy of $r_i* = (n_i / N)R$, where $R = \sum_{i=1}^{t} r_i$ is the sum of the Republican's expenditures across the toss-up states. This rule is called the **proportional rule**.

⌘ Example D

Assume there are three states with 7, 9, and 18 voters, and they are all toss-up states. If both the Democratic and Republican candidates choose strategies that maximize their expected popular vote (the proportional rule), and they have the same total resources $(D = R = 100)$, what resources should be allocated for each state?

Solution

$N = 7 + 9 + 18 = 34$, so the candidates should allocate $\frac{7}{34}(100) \approx 21$ for the first state, $\frac{9}{34}(100) \approx 26$ for the second state, and $\frac{18}{34}(100) \approx 53$ for the third state.

☞ Key idea

If a candidate wins more than 50% of the popular votes in a state, that candidate gets all of that state's electoral votes. If P_i is the probability that the Democrat wins more than 50% of the popular votes in toss-up state i, and that state has v_i electoral votes, then the **expected electoral vote (*EEV*)** of the Democratic candidate in toss-up states is

$$EEV_D = \sum_{i=1}^{t} v_i P_i,$$

where v_i is the number of electoral votes of toss-up state i, and P_i is the probability that the Democrat wins *more than* 50% of the popular votes in this state, which would give the Democrat *all* that state's electoral votes, v_i.

Section 12.7   Is There a Better Way to Elect a President?

☞ Key idea

The use of the Electoral College to determine the outcome of the presidential election has become controversial. Other election techniques, such as approval voting, could be considered instead.

Homework Help

Exercises 1 – 15
Carefully read Section 12.1 before responding to these exercises.

Exercises 16 – 23
Carefully read Section 12.2 before responding to these exercises.

Exercises 24 – 27
Carefully read Section 12.3 before responding to these exercises.

Exercises 28 – 34
Carefully read Section 12.4 before responding to these exercises.

Exercises 35 – 44
Carefully read Section 12.5 before responding to these exercises.

Exercises 45 – 53
Carefully read Section 12.6 before responding to these exercises.

Do You Know the Terms?

Cut out the following 29 flashcards to test yourself on Review Vocabulary. You can also find these flashcards at http://www.whfreeman.com/fapp7e.

Chapter 12 **Electing the President** **Approval voting**	**Chapter 12** **Electing the President** **Bandwagon effect**
Chapter 12 **Electing the President** **Condorcet candidate**	**Chapter 12** **Electing the President** **Dichotomous preferences**
Chapter 12 **Electing the President** **Discrete distribution of voters**	**Chapter 12** **Electing the President** **Dominant strategy**
Chapter 12 **Electing the President** **Electoral College**	**Chapter 12** **Electing the President** **Equilibrium position**

Voting for a candidate not on the basis of merit but, instead, because of the expectation that he or she will win.	Allows voters to vote for as many candidates as they like or find acceptable. Each candidate approved of receives one vote, and the candidate with the most approval votes wins.
Held by voters who divide the set of candidates into two subsets — a preferred subset and a nonpreferred subset — and are indifferent among all candidates in each subset.	A candidate who can defeat each of the other candidates in pairwise contests.
A strategy that is at least as good as, and sometimes better than, any other strategy.	A distribution in which voters are located at only certain positions along the left-right continuum.
A position is in equilibrium if no candidate has an incentive to depart from it unilaterally.	A body of 538 electors that selects a president.

Chapter 12 Electing the President **Expected electoral vote *(EEV)***	Chapter 12 Electing the President **Expected popular vote *(EPV)***
Chapter 12 Electing the President **Extended median**	Chapter 12 Electing the President **Global maximum**
Chapter 12 Electing the President **Local maximum**	Chapter 12 Electing the President **Maximin position**
Chapter 12 Electing the President **Mean $(\bar{l})$**	Chapter 12 Electing the President **Median (M)**

In toss-up states, the number of voters in each toss-up state, multiplied by the probability that that voter votes for the Democratic (or Republican) candidate, summed across all toss-up states.	In toss-up states, the number of electoral votes of each toss-up state, multiplied by the probability that the Democratic (or Republican) candidate wins more than 50% of the popular votes in that state, summed across all toss-up states.
A maximizing strategy from which *all* deviations (large or small) are nonoptimal.	The equilibrium position of two candidates when there is no median.
A position is maximin for a candidate if there is no other position that can guarantee a better outcome — more votes — whatever position another candidate adopts.	A maximizing strategy from which small deviations are nonoptimal but large deviations may be optimal.
The point on the horizontal axis of a voter distribution where half the voters have attitudes that lie to the left and half to the right.	A weighted average, wherein the positions of voters are weighted by the fraction of voters at that position.

Chapter 12 Electing the President **Median-voter theorem**	Chapter 12 Electing the President **Mode**
Chapter 12 Electing the President **1/3-separation obstacle**	Chapter 12 Electing the President **Poll assumption**
Chapter 12 Electing the President **Plurality voting**	Chapter 12 Electing the President **Proportional rule**
Chapter 12 Electing the President **Sincere voting**	Chapter 12 Electing the President **Spatial models**

A peak of a distribution. A distribution is unimodal if it has one peak and bimodal if it has two peaks.	In a two-candidate election with an odd number of voters, the median is the unique equilibrium position.
Voters adjust their sincere voting strategies, if necessary, to differentiate between the top two candidates — as revealed in the poll —by voting for the one they prefer.	An obstacle for the entry of a third candidate created if two previous entrants are sufficiently close together.
Presidential candidates allocate their resources to toss-up states according to their size. This allocation rule maximizes the expected popular vote of a candidate, given that his or her opponent adheres to it. It is a global maximum	Allows voters to vote for one candidate, and the candidate with the most votes wins.
The representation of candidate positions along a left – right continuum in order to determine the equilibrium or optimal positions of the candidates.	Voting for a favorite candidate, whatever his or her chances are of winning.

Chapter 12
Electing the President

Spoiler problem

Chapter 12
Electing the President

Strategic voting

Chapter 12
Electing the President

3/2's rule

Chapter 12
Electing the President

2/3-separation opportunity

Chapter 12
Electing the President

Voter distribution

Voting that is not sincere but nevertheless has a strategic purpose — namely, to elect an acceptable candidate if one's first choice is not viable.	Caused by a candidate who cannot win but "spoils" the election for a candidate who otherwise would win.
An opportunity for the entry of a third candidate created if two previous entrants are sufficiently far apart.	Presidential candidates allocate their resources to toss-up states according to the 3/2's power of their electoral votes. This allocation rule maximizes the expected electoral vote of a candidate, given that his or her opponent adheres to it. It is a local maximum.
	Gives the number (or percentage) of voters who have attitudes at each point along the left–right continuum, which can be represented by a curve. The distribution is symmetric if the curve to the left of the median is a mirror image of the curve to the right. It is skewed if the area under the curve is concentrated more on one side of the median than the other.

Practice Quiz

1. In a two-candidate election, assume the attitudes of the voters are symmetrically distributed with median M. If candidate A takes a position at M and candidate B takes a position to the left of M, then

    a. B gets the majority of the votes.
    b. A gets at least 50% of the votes.
    c. A and B each get 50% of the votes.

2. Find the median, M, for the discrete distribution of $n = 16$ voters at $k = 9$ different positions over the interval $[0,1]$ shown in the table below.

Position i	1	2	3	4	5	6	7	8	9
Location (l_i) of position i	0.1	0.2	0.3	0.4	0.5	0.6	0.7	0.8	0.9
Number of voters (n_i) at position i	2	2	3	0	2	1	5	1	0

    a. 0.7
    b. 0.4
    c. 0.5

3. Find the mean, $\bar{l}$, for the discrete distribution of $n = 16$ voters at $k = 9$ different positions over the interval $[0,1]$ shown in the table below.

Position i	1	2	3	4	5	6	7	8	9
Location (l_i) of position i	0.1	0.2	0.3	0.4	0.5	0.6	0.7	0.8	0.9
Number of voters (n_i) at position i	2	2	3	0	2	1	5	1	0

    a. 0.46
    b. 0.50
    c. 0.70

4. In a three-candidate race, candidates A and B take a position equidistant from the median, M, of a symmetric unimodal distribution and no more than $\frac{1}{3}$ of the voters lie between them. What position should candidate C take to win?

    a. C should take a position on either side of M.
    b. C should take a position at M.
    c. No position will result in a win for C.

5. An election procedure in which each voter votes for one candidate and the candidate with the most votes wins is called _____ .

    a. plurality voting.
    b. approval voting.
    c. Condorcet winner.

6. Fourteen voters list their voter preference for candidates A, B, and C according to the table below.

I.	5:	A	B	C
II.	3:	C	A	B
III.	6:	B	C	A

Which of the following is true?

a. The plurality winner is A.

b. The plurality winner is B.

c. The plurality winner is C.

7. Fourteen voters list their voter preference for candidates A, B, and C according to the table below.

I.	5:	A	B	C
II.	3:	C	A	B
III.	6:	B	C	A

Which of the following is true?

a. The Condorcet winner is Candidate B.

b. The Condorcet winner is Candidate C.

c. There is no Condorcet winner.

8. In a four-candidate race, 34 voters list their voter approval for candidates A, B, C and D according to the table below. The candidate that wins under approval voting is _____ .

	10	13	11
A	×	×	
B			×
C		×	×
D	×		×

a. Candidate A

b. Candidates B and D

c. Candidate C

9. Suppose there are four toss-up states with 2, 3, 3, and 4 electoral votes. If the Republican follows the optimal strategy of spending $60,000,000 in the proportion 2:3:3:4 and the Democrat spends in the proportion 0:3:3:4, the Republican receives on average an expected popular vote of _____ .

a. 5.45 votes

b. 6.55 votes

c. 10.0 votes

10. Suppose that states A, B, and C have, respectively, 4, 9, and 25 voters, and these are also their numbers of electoral votes. If each of these states is a toss-up state, the 3/2's rule says that the candidates should allocate their resources in the proportions

a. 4:9:25

b. 16:81:625

c. 8:27:125

Word Search

Refer to pages 463 – 464 of your text to obtain the Review Vocabulary. There are 26 hidden vocabulary words/expressions in the word search below. *2/3-separation opportunity*, *3/2's rule*, and *1/3-separation obstacle* were omitted from the word search. *Mode* and *Spatial Models* appear separately. It should be noted that spaces and hyphens are removed.

```
B S F T I I K O H O X I E P M Y C N A H E U Q O E O Y N
E T N O T J S T A Y L I E I Z F G E F L E K I H A G H C
V M D O V Z G G E I O I S O S O E I G V T R L S E F O N
M E D I A N P Y S C A S P A T I A L M O D E L S A W P G
E V L S G R F E D G R H E L U R L A N O I T R O P O R P
S P O I L E R P R O B L E M C Z M F S P C E E B F V Z L
E E R N O X T D A A G A E A O T M S G H H Z G C Y R R U
P S Z C B P S Z B T Q I I O S P K B U H O P D E V E N R
A A E E A E S G N I T O V C I G E T A R T S R F X W O A
N E D R L C M C E A I R Y S N T N C D A O I D N O C I L
E U W E M T N O I T I S O P N I M I X A M I E S T Q T I
R D T V A E L F Y S E U O V C N D Q C T O C E F C F I T
E P I O X D M Z L T L O C A L M A X I M U M H C M Y S Y
S D A T I P N A G D S N O I T U B I R T S I D R E T O V
T I P I M O T P G N I T O V L A V O R P P A L M S C P O
K T R N U P L R E T A D I D N A C T E C R O D N O C M T
L E E G M U P O H I N A I D E M D E D N E T X E E U U I
A M P O L L A S S U M P T I O N M R S C F B S C S F I N
O I S I R A E E G E L L O C L A R O T C E L E A S E R G
Y G E T A R T S T N A N I M O D S K I N R K R W H R B V
S R E T O V F O N O I T U B I R T S I D E T E R C S I D
L S K T F O E R M E R O E H T R E T O V N A I D E M L L
X W B I E T O V L A R O T C E L E D E T C E P X E A I E
I M E A N E E F E E B A N D W A G O N E F F E C T U J
C I L R V D M K O E M M I P N K E E E E S T I A H R Q O
U T D A O R O C M F N W Y W P Z B O L S E R Q D D T E O
M I B M I T M P N E C G M O G F E B E I F E P I N F L P
S E Z J Z N S C L G T L M C B M K R M I E A I Z S A E E
```

1. _____
2. _____
3. _____
4. _____
5. _____
6. _____
7. _____
8. _____
9. _____
10. _____
11. _____
12. _____
13. _____

14. _____
15. _____
16. _____
17. _____
18. _____
19. _____
20. _____
21. _____
22. _____
23. _____
24. _____
25. _____
26. _____

Chapter 13
Fair Division

Chapter Objectives

Check off these skills when you feel that you have mastered them.

☐ Describe the goal of a fair-division problem.

☐ Define the term "player."

☐ Use the adjusted winner procedure to determine the division of a set of objects among two players.

☐ Define the terms equitable, envy-free, and Pareto-optimal.

☐ Use the Knaster inheritance procedure to determine the division of a set of objects among more than two players.

☐ Use cake-division procedures to provide a fair allocation to two or more players.

☐ Use preference lists and the "bottom-up" approach to work out optimal strategies for two players taking turns to divide a collection of assets.

Guided Reading

Introduction

Fair-division problems arise in many situations, including divorce, inheritance, or the liquidation of a business. The problem is for the individuals involved, called the **players**, to devise a scheme for dividing an object or a set of objects in such a way that each of the players obtains a share that she considers fair. Such a scheme is called a **fair-division procedure**.

Section 13.1   The Adjusted Winner Procedure

☞ Key idea

In the **adjusted winner procedure** for two players, each of the players is given 100 points to distribute over the items that are to be divided. Each party is then initially given those items for which he or she placed more points than the other party. The steps are as follows.

- Each party distributes 100 points over the items in a way that reflects their relative worth to that party.
- Each item is initially given to the party that assigned it more points. Each party then assesses how many of his or her own points he or she has received. The party with the fewest points is now given each item on which both parties placed the same number of points.
- Let A denote the party with the higher point total and B be the other party. Determine the following for each item.

$$\frac{A\text{'s point value of the item}}{B\text{'s point value of the item}}$$

Transfer items from A to B, in order of increasing **point ratio**, until the point totals are equal. (The point at which equality is achieved may involve a fractional transfer of one item.)

ᕦ Example A

Suppose that Adam and Nadia place the following valuations on the three major assets, which will be divided up:

Asset	Adam	Nadia
House	40	60
Car	30	20
Boat	30	20

a)   Who gets each of the assets initially?

b)   How many points does each of the players get according to this allocation?

c)   Is any further exchange of property necessary in order to equalize the allocations?

Solution

a)   Since the person who bids the most points on an item is initially assigned that item, Nadia gets the house, while Adam gets the car and boat.

b)   They have 60 points each.

c)   No, both Adam and Nadia have received an equal number of points.

ᗕᔕ Example B

Suppose that Adam and Nadia place the following valuations on the three major assets, which will be divided up:

Asset	Adam	Nadia
House	30	60
Car	30	10
Boat	40	30

a)   Who gets each of the assets initially?

b)   How many points does each of the players get according to this allocation?

c)   What further exchange of property is necessary in order to equalize the allocations?

d)   After this final reallocation of property, how many points does each of the players receive?

Solution

a)   Nadia gets the house, while Adam gets the car and boat.

b)   Adam gets 70 points (30 for the car and 40 for the boat), while Nadia gets 60 for the house.

c)   Some of Adam's property has to be transferred to Nadia, since he has received more points initially. To determine how much, we first compute the fractions.

$$\text{Car: } \tfrac{30}{10} \quad \text{Boat: } \tfrac{40}{30}$$

We now transfer part of Adam's assets to Nadia as follows.

- Let x equal the fraction of the boat which Adam will retain.
- To equalize the number of points, we solve the following equation.

$$30 + 40x = 60 + 30(1-x)$$
$$30 + 40x = 60 + 30 - 30x$$
$$30 + 40x = 90 - 30x$$
$$70x = 60 \Rightarrow x = \tfrac{6}{7}$$

- Since Adam retains $\tfrac{6}{7}$ of the boat, he must transfer $1 - \tfrac{6}{7} = \tfrac{1}{7}$ of the boat to Nadia.

d)   Adam gets to keep the car, which is worth 30 points to him, he also retains a $\tfrac{6}{7}$ share in the boat, which is worth $40\left(\tfrac{6}{7}\right) = 34\tfrac{2}{7}$ points, giving him a total of $64\tfrac{2}{7}$ points. Nadia gets the house, worth 60 points to her, and a $\tfrac{1}{7}$ share in the boat, worth $30\left(\tfrac{1}{7}\right) = 4\tfrac{2}{7}$ points, totaling $64\tfrac{2}{7}$ points.

⚷ Key idea

The adjusted winner procedure satisfies three important properties. The allocation must be:

- **equitable:** Both players receive the same number of points.
- **envy-free:** Neither player would be happier with what the other received.
- **Pareto-optimal:** No other allocation, arrived at by any means, can make one party better off without making the other party worse off.

✎ Question 1

Suppose that Adam and Nadia place the following valuations on the three major assets, which will be divided up:

Asset	Adam	Nadia
House	30	40
Car	60	20
Boat	10	40

After this final reallocation of property, how many points does each of the players receive?

Answer

$68\tfrac{4}{7}$ points

Section 13.2 The Knaster Inheritance Procedure

🗝 Key idea

The adjusted winner procedure applies only when there are two heirs. With three or more, the **Knaster inheritance procedure** can be used.

Example C

Ali, Bara, and Amina inherit a house. Their respective evaluations of the house are $105,000, $90,000, and $120,000. Describe a fair division.

Solution

The highest bidder gets the asset. Since Amina values the house at $120,000, her share is $\frac{\$120{,}000}{3}$=$40,000 and pays $80,000 into a **kitty**. Ali and Bara view their fair shares as $\frac{\$105{,}000}{3}$= $35,000 and $\frac{\$90{,}000}{3}$= $30,000, respectively. After they take these amounts from the kitty, $15,000 remains. This sum is then split equally among the three. Thus, Amina gets the house, and pays $40,000 to Ali and $35,000 to Bara.

Example D

Ali, Bara, and Amina are heirs to an estate, which consists of a house, an antique car, and a boat (named the Baby Boo). Their evaluations of the items in the estate follow.

Asset	Ali	Bara	Amina
House	$120,000	$135,000	$100,000
Car	$40,000	$30,000	$22,000
Boat	$80,000	$60,000	$70,000

Describe a fair division.

Solution

It is possible to solve this problem by doing a calculation similar to the previous one for each of the three items, and then adding up the totals. However, an alternative method is to look at the entire estate. For example, Ali's bids indicate that he places a total value of $240,000 on the estate, which means that his share is $80,000. Similarly, Bara's estimate is $225,000, with her share being $75,000, while Amina's estimate is $192,000, entitling him to $64,000. Now Ali gets the car and boat having a total value (in his estimate) of $120,000, which is $40,000 over his fair share. Hence, he places $40,000 into a kitty. In the same way, Bara gets the house and places $60,000 in the kitty. Amina then removes her share of $64,000, which leaves $36,000 in the kitty. This amount is then divided equally among the three heirs.

Bara gets the house and pays $48,000 to Amina. Ali gets the antique car and the boat (Baby Boo) and pays $28,000 to Amina. Amina gets no items, but receives a total of $76,000 from Bara and Ali.

Section 13.3   Taking Turns

⌨ Key idea

Two people often split a collection of assets between them using the simplest and most natural of fair-division schemes: **taking turns**. If they each know the other's preferences among the assets, their best strategies may not be to choose their own most highly preferred asset first.

ᏻ Example E

Rachel and Kari attend a car auction and together win the bidding for a collection of four classic cars: a '59 Chevy (C), a '48 Packard (P), a '61 Austin-Healey (A) and a '47 Hudson (H). They will take turns to split the cars between them, with Rachel choosing first. Here are their preferences:

	Rachel's Ranking	Kari's Ranking
Best	C	P
Second best	P	H
Third best	H	C
Worst	A	A

If Rachel chooses the Chevy (her favorite) first, then Kari may choose her favorite, the Packard; Rachel loses her second best choice. Is there a way for Rachel to choose strategically to end up with both of her top two?

Solution

Rachel should choose the Packard first. Kari must now resort to choosing the Hudson (her second favorite). That leaves the Chevy available to Rachel in the second round.

⌨ Key idea

There is a general procedure for rational players, called the "bottom-up-strategy," for optimizing individual asset allocations when taking turns. It is based on two principles:

- You should never choose your least-preferred available option.
- You should never waste a choice on an option that will come to you automatically in a later round.

ᏻ Example F

What would the allocation of cars be in the last example if Rachel and Kari use the bottom-up strategy? Assume that Rachel goes first.

	Rachel's Ranking	Kari's Ranking
Best	C	P
Second best	P	H
Third best	H	C
Worst	A	A

Solution

Rachel: P   C
Kari: H   A

✎ Question 2

Who would get the worst choice if Kari went first?

Answer

Rachel

Section 13.4   Divide-and-Choose

☞ **Key idea**

A fair-division procedure known as **divide-and-choose** can be used if two people want to divide an object such as a cake or a piece of property. One of the people divides the object into two pieces, and the second person chooses either of the two pieces.

Section 13.5   Cake-Division Procedures: Proportionality

☞ **Key idea**

The divide-and-choose method cannot be used if there are more than two players. A **cake-division procedure** is a scheme that n players can use to divide a cake among themselves in a way which satisfies each player.

☞ **Key idea**

A cake-division scheme is said to be proportional if each player's strategy guarantees him a piece of size at least $\frac{1}{n}$ in his own estimation. It is envy-free if each player feels that no other player's piece is bigger than the one he has received.

☞ **Key idea**

When there are three players, the **lone-divider method** guarantees proportional shares.

∽ **Example G**

Suppose that players 1, 2, and 3 view a cake as follows.

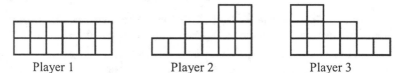

If player 1 cuts the cake into what she perceives as three equal pieces, draw three diagrams to show how each player will view the division.

Solution

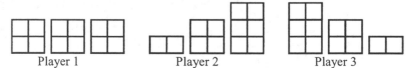

∽ **Example H**

If player 2 is the first to choose, which piece will he pick, and how much of the cake does he believe he is getting?

Solution

Player 2 believes that the pieces are cut unequally, with the right-most piece being the largest one. Thus Player 2 will select the right-most piece, which he believes is $\frac{1}{2}$ of the cake.

⁊ Example I

Suppose that Players 1, 2, and 3 view a cake as follows.

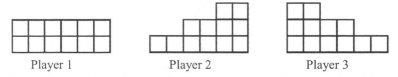

Player 1   Player 2   Player 3

a)   Which of the two remaining pieces will Player 3 choose, and how much of the cake does she believe she is getting?

b)   Which piece will be left for Player 1, and how much of the cake does she believe she is getting?

Solution

a) Player 3 believes that the pieces are cut unequally, with the left-most piece being the largest. Player 3 will choose the left-most piece, which she believes is $\frac{1}{2}$ of the cake.

b) Player 1 believes that all three pieces are of equal size. The middle piece will remain for player 1, and she believes that it is $\frac{1}{3}$ of the cake.

⊶ Key idea

The **last-diminisher method** is a more complicated procedure, which guarantees proportional shares with any number of players. Like the lone-divider method, it is proportional but not envy-free.

Section 13.6   Cake-Division Procedures: The Problem of Envy

⊶ Key idea

Envy-free cake-division schemes are still more complicated. The **Selfridge-Conway procedure** solves this problem for the case of three players. For more than three players, a scheme which involves a **trimming procedure** has been developed. In it, proportions of the cake are successively allocated in an envy-free fashion, with the remaining portions diminishing in size. Eventually, the remainder of the cake is so small that it will not affect the perception of each player that he or she has obtained the largest piece. If players cannot find a way to make an item fairly divisible, there may be no alternative but to sell it and share the proceeds equally. The procedure is as follows.

       Step 1   Player 1 cuts the cake into three pieces he considers to be the same size. He hands the three pieces to Player 2.

       Step 2   Player 2 trims at most one of the three pieces so as to create at least a two-way tie for largest. Setting the trimmings aside, Player 2 hands the three pieces (one of which may have been trimmed) to Player 3.

       Step 3   Player 3 now chooses, from among the three pieces, one that he considers to be at least tied for largest.

       Step 4   Player 2 next chooses, from the two remaining pieces, one that she considers to be at least tied for largest, with the proviso that if she trimmed a piece in Step 2, and Player 3 did not choose this piece, then she must now choose it.

       Step 5   Player 1 receives the remaining piece.

ᕰᕫ **Example J**

Suppose that Players 1, 2, and 3 view a cake as follows.

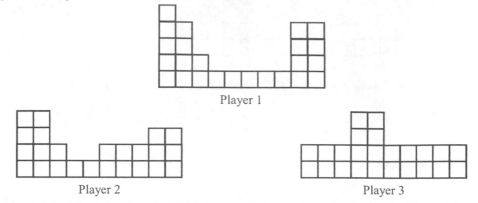

Provide a total of three drawings that show how each player views the division by Player 1 into three pieces he or she considers to be the same size or value. Label the pieces *A*, *B*, and *C*. Assume the cuts correspond to vertical lines. Identify pieces that are acceptable to Players 2 and 3.

Solution

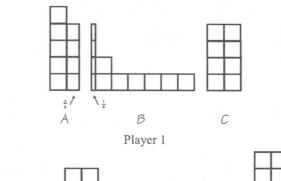

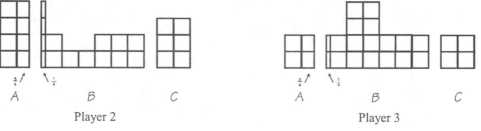

Since an acceptable piece would be at least of size $\frac{24}{3} = 8$, Player 2 would find piece *B* acceptable. Player 3 also would find piece *B* acceptable.

✏ **Question 3**

If Player 2 did the cutting, how many pieces would be acceptable to Players 1 and 3?

Answer

Two for Player 1 and 1 for Player 3.

Homework Help

Exercises 1 – 6
Carefully read Section 13.1 before responding to these exercises. Remember when calculating point ratios you calculate them relative to the person that has more total points in order to share with the other person. In Exercises 5 and 6, answers will vary.

Exercise 7
Carefully read Section 13.1 and pay special attention to the theorem on page 480.

Exercises 8 – 12
Carefully read Section 13.2 before responding to these exercises. In Exercise 10, answers will vary.

Exercises 13 – 18
Carefully read Section 13.3 before responding to these exercises. The following may be helpful.

Exercise 13

Bob: _____   _____   _____

Carol: _____   _____   _____

Exercise 14

Carol: _____   _____   _____

Bob: _____   _____   _____

Exercise 15

Mark: _____   _____   _____

Fred: _____   _____   _____

Exercise 16

Fred: _____   _____   _____

Mark: _____   _____   _____

Exercise 17

Donald: _____   _____   _____

Ivana: _____   _____

Exercise 18

Ivana: _____   _____   _____

Donald: _____   _____

Exercises 19 – 22
Carefully read Section 13.4 before responding to these exercises.

Exercises 23 – 28
Carefully read Section 13.4 before responding to these exercises.

Exercises 29 – 30
Carefully read Section 13.5 before responding to these exercises.

Do You Know the Terms?

Cut out the following 16 flashcards to test yourself on Review Vocabulary. You can also find these flashcards at http://www.whfreeman.com/fapp7e.

Chapter 13 Fair Division **Adjusted winner procedure**	Chapter 13 Fair Division **Bottom-up strategy**
Chapter 13 Fair Division **Cake-division procedure**	Chapter 13 Fair Division **Convention of the Law of the Sea**
Chapter 13 Fair Division **Divide-and-choose**	Chapter 13 Fair Division **Envy-free**
Chapter 13 Fair Division **Equitable**	Chapter 13 Fair Division **Knaster inheritance procedure**

A bottom-up strategy is a strategy under an alternating procedure in which sophisticated choices are determined by working backwards.	The allocation resulting from this procedure is equitable, envy-free, and Pareto-optimal. It requires no cash from either player, but one of the objects may have to be divided or shared by the two players.
An agreement based on divide-and-choose that protects the interests of developing countries in mining operations under the sea.	Such procedures involve finding allocations of a single object that is finely divisible, as opposed to the situation encountered with either the adjusted winner procedure or Knaster's procedure. Each player has a strategy that will guarantee that player a piece with which he or she is "satisfied," even in the face of collusion by the others.
A fair-division procedure is said to be envy-free if each player has a strategy that can guarantee him or her a share of whatever is being divided that is, in the eyes of that player, at least as large (or at least as desirable) as that received by any other player, no matter what the other players do.	A fair-division procedure for dividing an object or several objects between two players. This method produces an allocation that is both proportional and envy-free (the two being equivalent when there are only two players).
A fair-division procedure for any number of parties that begins by having each player (independently) assign a dollar value (a "bid") to the item or items to be divided so as to reflect the absolute worth of each object to that player. It never requires the dividing or sharing of an object, but it may require that the players have a large amount of cash on hand.	An allocation (resulting from a fair-division procedure like adjusted winner) is said to be equitable if each player believes he or she received the same fractional part of the total value

Chapter 13
Fair Division

Last-diminisher method

Chapter 13
Fair Division

Lone-divider method

Chapter 13
Fair Division

Pareto-optimal

Chapter 13
Fair Division

Point ratio

Chapter 13
Fair Division

Preference lists

Chapter 13
Fair Division

Proportional

Chapter 13
Fair Division

Selfridge-Conway envy-free procedure

Chapter 13
Fair Division

Taking turns

A cake-division procedure introduced by Hugo Steinhaus in the 1940s. It works only for three players and produces an allocation that is proportional but not, in general, envy-free.	A cake-division procedure introduced by Stefan Banach and Bronislaw Knaster in the 1940s. It works for any number of players and produces an allocation that is proportional but not, in general, envy-free.
The fraction in which the numerator is the number of points in one party placed on an object and the denominator is the number of points the other party placed on the object.	When no other allocation, achieved by any means whatsoever, can make any one player better off without making some other player worse off.
A fair-division procedure is said to be proportional if each of n players has a strategy that can guarantee that player a share of whatever is being divided that he or she considers to be at least $1/n$ of the whole in size or value.	Rankings of the items to be allocated, from best to worst, by each of the participants.
A fair-division procedure in which two or more parties alternate selecting objects.	A cake-division procedure introduced independently by John Selfridge and John Conway around 1960. It works only for three players but produces an allocation that is envy-free (as well as proportional).

Practice Quiz

1. In a fair-division procedure, the goal is for each person to receive
    a. an identical portion.
    b. what is perceived as an identical portion.
    c. what is perceived as an acceptable portion.

2. Using the adjusted winner procedure to divide property between two people, each person always receives
    a. exactly 50 points of value.
    b. at least 50 points of value.
    c. no more than 50 points of value.

3. Amina and Ali must make a fair division of three cars. They assign points to the cars, and use the adjusted winner procedure. Which car is divided between Amina and Ali?

	Amina	Ali
Red car	40	20
White car	25	30
Blue car	35	50

    a. Red car
    b. White car
    c. Blue car

4. Bara and Jenna must make a fair division of three art items. They assign points to the items, and use the adjusted winner procedure. Who ends up with the Sculpture?

	Bara	Jenna
Painting	35	50
Sculpture	40	35
Tapestry	25	15

    a. Bara
    b. Jenna
    c. Bara and Jenna share the Sculpture

5. Alex and Kari use the Knaster inheritance procedure to fairly divide a coin collection. Alex bids $700 and Kari bids $860. What is the outcome?
    a. Kari gets the coins and pays Alex $430.
    b. Kari gets the coins and pays Alex $390.
    c. Kari gets the coins and pays Alex $160.

6. Four children bid on two objects. Using the Knaster inheritance procedure, who ends up with the most cash money?

	Caleb	Quinn	Nadia	Adam
House	$81,000	$75,000	$82,000	$78,000
Car	$12,000	$11,000	$10,000	$13,000

   a. Caleb

   b. Quinn

   c. Nadia

7. Four children bid on two objects. Using the Knaster inheritance procedure, what does Adam do?

	Caleb	Quinn	Nadia	Adam
House	$81,000	$75,000	$82,000	$78,000
Car	$12,000	$11,000	$10,000	$13,000

   a. takes car and takes cash

   b. takes car and pays cash

   c. takes car only

8. Suppose John and Dean use the bottom-up strategy, taking turns to divide several items, ranked in order as shown below. If John goes first, what is his first choice?

	John's ranking	Dean's Ranking
1st choice	Clock	Radio
2nd choice	Radio	Phone
3rd choice	Toaster	Toaster
4th choice	Phone	Clock

   a. Clock

   b. Radio

   c. Toaster

9. Suppose seven people will share a cake using the last-diminisher method. To begin, Matthew cuts a piece and passes it to Dani. Dani trims the piece and passes it on to each of the remaining five people, but no one else trims the piece. Then

   a. Matthew gets this piece.

   b. Dani gets this piece.

   c. the last person who is handed the piece keeps it.

10. Which of the following procedures is envy-free?

   a. Lone-divider

   b. Selfridge-Conway

   c. Last-diminisher

Word Search

Refer to pages 493 – 494 of your text to obtain the Review Vocabulary.   There are 15 hidden vocabulary words/expressions in the word search below.   *Selfridge-Conway envy-free procedure* does not appear in the word search.   It should be noted that spaces are removed as well as hyphens.

```
V T N A Z A L R T C A O S E P O E J J A O O R J B E Y Q I L
H T E K I E Z W M V I K M C Z A N P P S O G Z N D Z K Q G D
E O T L P S N O I I S A M I Q E E A U O N H M X P E O L W D
V I T O R Y P R S A E Z S S H E Q I P T D R L Q T M S H T C
R C C P I Z S T G J I C F T O D G S I Y T C U W C C C M E T
E C H O R M A Y P N K X R G F O O R L K M A F T S W A D P T
E O R E N Z N E M Y N M T E O V F X C D S H E E G Q N I O W
A L E C D V P D P I D H I M N E L E Y Y E R Z R E N D C D P
O P E R U D E C O R P R E N N I W D E T S U J D A O I A R V
M R I H L C C N L C I X G T O B S O O X Y E H C H L E K D C
E E J T Q T L N T C P E O L T C V R N A L N H T R I E E A S
Z F L A S T D I M I N I S H E R M E T H O D E S O E C D M T
E E M S E F S C E Q O Y G E T A R T S P U M O T T O B I U C
E R U D E C O R P E C N A T I R E H N I R E T S A N K V E T
V E H H Z N N N G S E S O O H C D N A E D I V I D J D I L G
N N L M A S I N H L G R G F S Z C E D X O I X E N Z R S B E
F C R P Z O R Z A R F V I Q T E S I T R O R P X B W L I A S
S E N B K P S K C A Q P M S A H V S O Y D T A I T N M O T S
Q L I E F E O A F M E R O X H I E V Y P E N R D L U E N I A
F I D E S N S T W I V E O I D W S L D Q N R E T A I C P U H
T S G L Y S E E R F Y V N E F G G R A V F E T Y O H Z R Q E
D T O P S R N F S A R M N L P D S S C W R B O X S E Y O E R
O S N I N Q I E M W U O R L A N O I T R O P O R P S P C T W
M I E C T R I O A F L A L M M L O S G H B F P D O E S E W I
F I V J A K E Z H F E G I N H D H F Y O S T T I W I G D E B
L P R H S R D S S J E E B D D K Q F Q K I E I H O E L U S U
A P D L T T T V F Z N E E H U X Z T I H Z E M K E L R R T N
O X V F H C K K U P O I P O I N T R A T I O A E P S Q E Q H
Q P H C T A H C R E G L H E E B W M W T N O L S J A E C R I
H E R E A N E D L M Y P C M E E S E E H E E S E C E A A K N
```

1. _____   9. _____

2. _____   10. _____

3. _____   11. _____

4. _____   12. _____

5. _____   13. _____

6. _____   14. _____

7. _____   15. _____

8. _____

Chapter 14
Apportionment

Chapter Objectives

Check off these skills when you feel that you have mastered them.

☐ State the apportionment problem.

☐ Explain the difference between quota and apportionment.

☐ State the quota condition and be able to tell which apportionment methods satisfy it and which do not.

☐ Do the same for the house monotone and population monotone conditions.

☐ Know that some methods have bias in favor of large or small states.

☐ Recognize the difference in computing quotas between the Hamilton method and divisor methods.

☐ Calculate the apportionment of seats in a representative body when the individual population sizes and number of seats are given, using the methods of Hamilton, Jefferson, Webster, and Hill-Huntington.

☐ Be able to give at least three reasons to support the claim that Webster's method is the "best" apportionment method.

☐ Calculate the critical divisor for each state.

☐ Determine an apportionment using the method of critical multipliers.

☐ Explain why the Jefferson method and the method of critical multipliers do not satisfy the quota condition.

Guided Reading

Introduction

In many situations, a fixed number of places must be divided among several groups, in a way that is proportional to the size of each of the groups. A prime example of this is the division of the 435 seats in the U.S. House of Representatives among the 50 states. A problem arises because the exact allocations will usually involve fractional seats, which are not allowed. Various methods have been proposed to round the fraction to reach the total of 435, and four different methods have been used in the apportionment of the House during the past 200 years.

Section 14.1   The Apportionment Problem

⌐ Key idea

The **apportionment problem** is to round a set of fractions so that their sum is a fixed number. An **apportionment method** is a systematic procedure that solves the apportionment problem.

⌐ Key idea

The **standard divisor** is obtained by dividing the total population by the house size. A state's **quota** is obtained by dividing its population by the average district population. It represents the exact share the state is entitled to. However, since the quota is usually not a whole number, an apportionment method must be used to change each of the fractions into an integer.

⌐ Key idea

Two notations are used in this chapter for rounding.

$\lfloor q \rfloor$ means round down to the integer value $\lceil q \rceil$ means round up to the integer value

If q is an integer value, no rounding occurs.

⌒ Example A

Suppose that a country consists of three states, A, B, and C, with populations 11,000, 17,500, and 21,500, respectively. If the congress in this country has 10 seats, find the population of the average congressional district and the quota of each of the three states.

Solution

The total population is 50,000 and there are 10 districts, so that the average size is $\dfrac{50,000}{10} = 5000$.

We find each state's quota by dividing its population by 5000, the size of the average congressional district. Thus, A's quota is $\dfrac{11,000}{5000} = 2.2$, B's quota is $\dfrac{17,500}{5000} = 3.5$, and C's is $\dfrac{21,500}{5000} = 4.3$.

⌒ Example B

States A, B, and C have populations of 11,000, 17,500, and 21,500, respectively. If their country's congress has 10 seats, which of the following assignments of seats fulfills the conditions of an apportionment method?
a) A – 3 seats, B – 3 seats, C – 4 seats.
b) A – 2 seats, B – 3 seats, C – 5 seats.
c) A – 2 seats, B – 3 seats, C – 4 seats.
d) A – 2 seats, B – 4 seats, C – 4 seats.

Solution

a), b), and d) are legitimate apportionments, while c) is not. In c), the total number of seats allocated is 9, which is short of the house size of 10. The other three apportionments allocate 10 seats.

Section 14.2   The Hamilton Method

◦→ Key idea

The **lower quota** for a state is obtained by rounding its quota *down* to the nearest integer, while to obtain the **upper quota** we round the quota *up*.

◦→ Key idea

The **Hamilton method** first assigns each state its lower quota, and then distributes any remaining seats to the states having the largest fractional parts.

↷ Example C

Consider the country previously listed, in which state *A* has a population of 11,000, state *B*, 17,500, and state *C*, 21,500.   How would Hamilton allocate the 10 seats of the house?

Solution

The quotas are: *A* – 2.2, *B* – 3.5, *C* – 4.3.   Each state initially receives its lower quota, so that *A* receives two seats, *B* three, and *C* four.   The one remaining seat now goes to the state with the largest fractional part, which is *B*.   Hence, *B* also gets four seats.

↷ Example D

Enrollments for four mathematics courses are as follows.

Algebra–62

Geometry–52

Trigonometry–38

Calculus–28

Ten mathematics sections will be scheduled.   How many sections of each of these four courses will be allocated by the Hamilton method?

Solution

Since there are 180 students and 10 sections, the average size of a section will be $\dfrac{180}{10} = 18$.

$$\text{Quota for Algebra} = \frac{\text{population}}{\text{divisor}} = \frac{62}{18} = 3.44$$

$$\text{Quota for Geometry} = \frac{\text{population}}{\text{divisor}} = \frac{52}{18} = 2.89$$

$$\text{Quota for Trigonometry} = \frac{\text{population}}{\text{divisor}} = \frac{38}{18} = 2.11$$

$$\text{Quota for Calculus} = \frac{\text{population}}{\text{divisor}} = \frac{28}{18} = 1.56$$

Distributing the lower quotas, Algebra gets three sections, Geometry and Trigonometry get two each, while Calculus gets one.   The Hamilton method now assigns the remaining two sections to the courses having the largest fractional parts.   Thus, Geometry gets the first section, and Calculus gets the second.   There will be three sections of Algebra and Geometry, and two sections of Trigonometry and Calculus.

✏ Question 1

Four classes need to assign 12 boxes of supplies according to their size using the Hamilton method. How many boxes does each class receive?

$$A–28 \quad B–17 \quad C–24 \quad D–31$$

Answer

A will receive 3, B will receive 2, C will receive 3, and D will receive 4.

⚷ Key idea

The **Alabama paradox** occurs if a state loses one or more seats when the number of seats in the house is increased. The **population paradox** occurs if a state loses at least one seat, even though its population increases, while another state gains at least one seat, even though its population decreases. The Hamilton method is susceptible to both of these paradoxes.

Section 14.3   Divisor Methods

⚷ Key idea

A **divisor method** determines each state's apportionment by dividing its population by a common divisor d and rounding the resulting quotient. If the total number of seats allocated, using the chosen divisor, does not equal the house size, a larger or smaller divisor must be chosen. Finding a decisive divisor for a method of apportionment depends on how the fractions are rounded.

⚷ Key idea

In the **Jefferson method** the fractions are all rounded *down*.

ᴄᴀ Example E

In the example of the three states (A: 11,000, B: 17,500, C: 21,500), would a divisor of 5000 be decisive in apportioning a house of 10 seats, according to the Jefferson method?

Solution

With a divisor of 5000, the quotas for A, B, and C are 2.2, 3.5 and 4.3, respectively. Since all fractions are rounded down in the Jefferson method, A would receive two seats, B three, and C four, for a total of nine, one fewer than the number to be allocated. Hence, 5000 is not a decisive divisor.

ᴄᴀ Example F

In the example of the three states (A: 11,000, B: 17,500, C: 21,500), would a divisor of 4350 be decisive for the Jefferson method?

Solution

A's quota is $\dfrac{11,000}{4350} = 2.53$, B's quota is $\dfrac{17,500}{4350} = 4.02$, and C's is $\dfrac{21,500}{4350} = 4.94$. Rounding down gives A two seats, and B and C four each, for a total of ten. Thus the answer is yes.

✐ Question 2

Four classes need to assign 12 boxes of supplies according to their size. Would a divisor of 9 be decisive for the Jefferson method?

$$A–28 \quad B–17 \quad C–24 \quad D–31$$

Answer

No.

☞ Key idea

Finding a decisive divisor by trial and error can be quite tedious. A systematic method for doing so involves calculating the critical divisor for each state.

⌇ Example G

Consider, again, the case of the four mathematics courses and their enrollment totals:

Algebra–62

Geometry–52

Trigonometry–38

Calculus–28

Ten mathematics sections will be scheduled. Use the method of critical divisors to determine how many sections of each of these courses will be allocated by the Jefferson method.

Solution

In the Jefferson method, fractions are rounded down. Recall the following quotas.

Quota for Algebra = 3.44

Quota for Geometry = 2.89

Quota for Trigonometry = 2.11

Quota for Calculus = 1.56

The tentative allocation of sections is three to Algebra, two each to Geometry and Trigonometry, and one to Calculus. To determine which courses get the two remaining sections, we compute the critical divisors for each course. We obtain this for Algebra by dividing Algebra's enrollment, 62, by 4, which is one more than its tentative allocation.

$$\text{Critical divisor for Algebra} = \frac{62}{4} = 15.5$$

$$\text{Critical divisor for Geometry} = \frac{52}{3} = 17.33$$

$$\text{Critical divisor for Trigonometry} = \frac{38}{3} = 12.67$$

$$\text{Critical divisor for Calculus} = \frac{28}{2} = 14$$

Since Geometry's critical divisor is largest, it receives the next section. Before proceeding to allocate the next section, we recalculate Geometry's critical divisor, since it now has 3 sections. Its new divisor is $\frac{52}{4} = 13$. At this point, Algebra has the largest critical divisor, and it receives the last section. Thus, Algebra gets four sections, Geometry gets three, Trigonometry gets two and Calculus gets one.

✎ Question 3

Four classes need to assign 12 boxes of supplies according to their size. Use the method of critical divisors to determine how many boxes each class will be allocated by the Jefferson method.

$$A-28 \quad B-17 \quad C-24 \quad D-31$$

Answer

A will receive 3, *B* will receive 2, *C* will receive 3, and *D* will receive 4.

⚷ Key idea

The **Webster method** is also a divisor method, in which fractions greater than or equal to 0.5 $\left(\frac{1}{2}\right)$ are rounded up, while those less than 0.5 $\left(\frac{1}{2}\right)$ are rounded down.

↝ Example H

Ten mathematics sections will be scheduled for four groups of students:

Algebra–62

Geometry–52

Trigonometry–38

Calculus–28

Use the method of critical divisors to determine how many sections of each of these courses will be allocated by the Webster method.

Solution

As we have seen, the quotas for the four subjects are 3.44, 2.89, 2.11, and 1.56, respectively. Now, however, we round the fractions in the normal way to obtain the Webster apportionment. Hence, Algebra, the fractional part of whose quota is less than 0.5, receives three seats in the tentative apportionment. Trigonometry's allocation is also rounded down to two, while Geometry and Calculus, both of whose fractional parts are greater than 0.5, have their allocations rounded up, Geometry to three and Calculus to two. Note: If the sum of the tentative allocations had been less than ten, we would have to find a smaller divisor. In either case, the method of critical divisors that we introduced in connection with the Jefferson method can be suitably modified to give the correct Webster apportionment. Trial and error can also be used to find an appropriate divisor. Algebra and Geometry get three sections each, while Trigonometry and Calculus get two.

✎ Question 4

Four classes need to assign 12 boxes of supplies according to their size. Use the method of critical divisors to determine how many boxes each class will be allocated by the Webster method.

$$A-28 \quad B-17 \quad C-24 \quad D-31$$

Answer

A will receive 3, *B* will receive 2, *C* will receive 3, and *D* will receive 4.

⚷ Key idea

In the Jefferson method, *all* fractions are rounded down, while in the Webster method, the cutoff point for rounding is 0.5. In the Hill-Huntington method, the cutoff point depends upon the size of the apportionment. If a state's quota is n seats, then its cutoff point is the geometric mean of n and $n+1$, which is $\sqrt{n(n+1)}$. For example, if a state's quota is between 4 and 5, then $n=4$, so that the cutoff point is $\sqrt{4(5)} \approx 4.472$. Thus, if a state's quota is 4.37, the state would get just 4 seats, since $4.37 < 4.472$. On the other hand, if its quota is 4.48, which is greater than the cutoff point of 4.472, it would get five seats.

Ꮾᐟ **Example I**
Ten mathematics sections will be scheduled for four groups of students:

Algebra–62

Geometry–52

Trigonometry–38

Calculus–28

Use the method of critical divisors to determine how many sections of each of these courses will be allocated by the Hill-Huntington method.

Solution
The quotas for the four courses are 3.44, 2.89, 2.11 and 1.56, respectively. We now compute the cutoff points for rounding up or down. For Algebra, the cutoff point is $\sqrt{3(4)} \approx 3.46$. Since Algebra's quota is 3.44, which is less than the cutoff point, Algebra's tentative allocation is rounded *down*. The cutoff points for Geometry and Trigonometry are both $\sqrt{2(3)} \approx 2.45$, so Geometry gets a third section, but Trigonometry does not. Finally, Calculus' cutoff point is $\sqrt{1(2)} \approx 1.41$; so Calculus also gets an extra section, for a total of two. Algebra and Geometry each get three seats, while Trigonometry and Calculus get two each. Since the sum of the tentative allocations is 10, the apportionment process is completed.

✎ **Question 5**
Four classes need to assign 12 boxes of supplies according to their size. Use the method of critical divisors to determine how many boxes each class will be allocated by the Hill-Huntington method.

A–28

B–17

C–24

D–31

Answer
A will receive 3, *B* will receive 2, *C* will receive 3, and *D* will receive 4.

Section 14.4   Which Divisor Method is Best?

ᐅ Key idea
All of the apportionment methods attempt to minimize inequities between the states, although they all use different criteria to measure the inequities.

ᐅ Key idea
The Webster method minimizes the inequity in the **absolute difference** in **representative shares**.

ᐅ Key idea
The Hill-Huntington method minimizes the relative difference in **representative shares** or **district populations**.

ᐅ Key idea
Given positive integers A and B with $A > B$, the **absolute difference** is $A - B$. The **relative difference** is $\frac{A-B}{B} \times 100\%$.

ᏏᏊ Example J

What is the relative difference between the numbers 8 and 13?

Solution

Because $A > B$, $A = 13$ and $B = 8$. The relative difference is $\dfrac{13-8}{8} \times 100\% = 0.625 \times 100\% = 62.5\%$.

ᏏᏊ Example K

In the 1990 census, Alabama's population was 4,040,587, and Arizona's was 3,665,228. Alabama was apportioned 7 congressional seats, and Arizona received 6. What is the size of the average congressional district in each of these two states? Which of these two states is more favored by this apportionment?

Solution

Alabama, whose average district population is 577,227, is favored over Arizona, whose average district population is 610,871. Each of these averages is found by dividing the state's population by the number of seats.

ᏏᏊ Example L

What is the relative difference in average district population between Alabama (577,227) and Arizona (610,871)?

Solution

The relative difference is $\dfrac{610,871-577,227}{577,227} \times 100\% \approx 0.058 \times 100\% = 5.8\%$.

Homework Help

Exercises 1 – 4
Carefully read Section 14.1 before responding to these exercises.

Exercises 5 – 13
Carefully read Section 14.2 before responding to these exercises.

Exercises 14 – 31
Carefully read Section 14.3 before responding to these exercises.

Exercises 32 – 44
Carefully read Section 14.4 before responding to these exercises.

Do You Know the Terms?

Cut out the following 27 flashcards to test yourself on Review Vocabulary. You can also find these flashcards at http://www.whfreeman.com/fapp7e.

Chapter 14 Apportionment $\lfloor q \rfloor$	Chapter 14 Apportionment $\lceil q \rceil$
Chapter 14 Apportionment **Absolute difference**	Chapter 14 Apportionment **Adjusted quota**
Chapter 14 Apportionment **Alabama paradox**	Chapter 14 Apportionment **Apportionment method**
Chapter 14 Apportionment **Apportionment problem**	Chapter 14 Apportionment **Critical divisor**

The result of rounding a number q up to the next integer.

The result of rounding a number q down.

The result of dividing a state's quota by a divisor other than the standard divisor. The purpose of adjusting the quotas is to correct a failure of the rounded quotas to sum to the house size.

The result of subtracting the smaller number from the larger.

A systematic way of computing solutions of apportionment problems.

A state loses a representative solely because the size of the House is increased. This paradox is possible with the Hamilton method but not with divisor methods.

The number closest to the standard divisor that can be used as a divisor of a state's population to obtain a new tentative apportionment for the state.

To round a list of fractions to whole numbers in a way that preserves the sum of the original fractions.

Chapter 14
Apportionment

Critical divisor for the Jefferson method causing change in tentative apportionment

Chapter 14
Apportionment

Critical divisor for the Webster method causing change in tentative apportionment

Chapter 14
Apportionment

Critical divisor for the Hill-Huntington method causing change in tentative apportionment

Chapter 14
Apportionment

District population

Chapter 14
Apportionment

Divisor method

Chapter 14
Apportionment

Geometric mean

Chapter 14
Apportionment

Hamilton method

Chapter 14
Apportionment

Hill – Huntington method

p stands for the state's population, and n is its tentative apportionment. to increase: $\dfrac{p}{n+0.5}$ to decrease: $\dfrac{p}{n-0.5}$	p stands for the state's population, and n is its tentative apportionment. to increase: $\dfrac{p}{n+1}$ to decrease: not necessary
A state's population divided by its apportionment.	p stands for the state's population, and n is its tentative apportionment. to increase: $\dfrac{p}{\sqrt{n(n+1)}}$ to decrease: $\dfrac{p}{\sqrt{n(n-1)}}$
For positive numbers A and B, the geometric mean is defined to be $\sqrt{AB}$.	One of many apportionment methods in which the apportionments are determined by dividing the population of each state by a common divisor to obtain adjusted quotas. The apportionments are calculated by rounding the adjusted quotas.
A divisor method that minimizes relative differences in both representative shares and district populations.	An apportionment method that assigns to each state either its lower quota or its upper quota. The states that receive their upper quotas are those whose quotas have the largest fractional parts.

Chapter 14 Apportionment **Jefferson method**	**Chapter 14** Apportionment **Lower quota**
Chapter 14 Apportionment **Population paradox**	**Chapter 14** Apportionment **Quota**
Chapter 14 Apportionment **Quota condition**	**Chapter 14** Apportionment **Relative difference**
Chapter 14 Apportionment **Representative share**	**Chapter 14** Apportionment **Standard divisor**

The integer part $\lfloor q_i \rfloor$ of a state's quota q_i.	A divisor method invented by Thomas Jefferson, based on rounding all fractions down. Thus, if u_i is the adjusted quota of state i, the state's apportionment is $\lfloor u_i \rfloor$.
The quotient p/s of a state's population divided by the standard divisor. The quota is the number of seats a state would receive if fractional seats could be awarded.	A situation is which state A gains population and loses a congressional seat, while state B loses population (or increases population proportionally less than state A) and gains a seat. This paradox is possible with all apportionment methods *except* divisor methods.
Subtracting the smaller number from the larger of two positive numbers, and expressing the result as a percentage of the smaller number.	A requirement that an apportionment method should assign to each state either its lower quota or its upper quota in every situation. The Hamilton method satisfies this condition, but none of the divisor methods do.
The ratio p/h of the total population p to the house size h. In a congressional apportionment problem, the standard divisor represents the average district population.	A state's representative share is the state's apportionment divided by its population. It is intended to represent the amount of influence a citizen of that state would have on his or her representative.

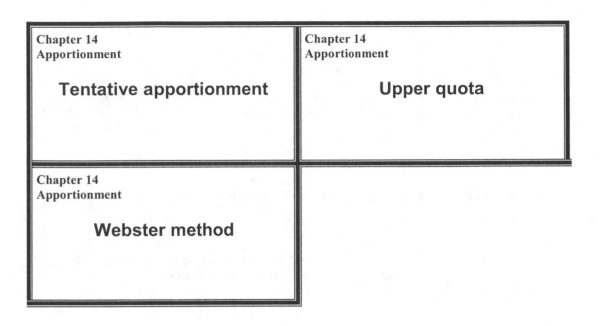

Chapter 14
Apportionment

Tentative apportionment

Chapter 14
Apportionment

Upper quota

Chapter 14
Apportionment

Webster method

The result of rounding a state's quota *up* to a whole number. A state whose quota is q has an upper quota equal to $\lceil q \rceil$.

The result of rounding a state's quota or adjusted quota to obtain a whole number.

A divisor method based on rounding fractions the usual way.

Practice Quiz

1. A county is divided into three districts with populations: Central, 3100; Western, 3500; Eastern, 1700. There are six seats on the county council to be apportioned. What is the quota for the Eastern district?

    a. less than 1

    b. 1

    c. more than 1

2. The Hamilton method of apportionment can display

    a. the population paradox, but not the Alabama paradox.

    b. the Alabama paradox, but not the population paradox.

    c. both the Alabama paradox and the population paradox.

3. The Jefferson method of apportionment
    a. is a divisor method.
    b. satisfies the quota condition.
    c. is biased in favor of smaller states.

4. The Webster method of apportionment

    a. is susceptible to the Alabama paradox.

    b. favors smaller states.

    c. can have ties.

5. A county is divided into three districts with populations Central, 3100; Western, 3500; Eastern, 1700. There are nine seats on the school board to be apportioned. What is the apportionment for the Eastern district using the Hamilton method?

    a. 1

    b. 2

    c. 3

6. A county is divided into three districts with populations: Central, 3100; Western, 3500; Eastern, 1700. There are nine seats on the school board to be apportioned. What is the apportionment for the Eastern district using the Jefferson method?

    a. 0

    b. 1

    c. 2

7. A county is divided into three districts with populations: Central, 3100; Western, 3500; Eastern, 1700. There are nine seats on the school board to be apportioned. What is the apportionment for the Eastern district using the Webster method?

    a. 1

    b. 2

    c. 3

8. A county is divided into three districts with populations: Central, 3100; Northern, 1900; Southern, 2800. There are five seats on the zoning board to be apportioned. What is the apportionment for the Southern district using the Hill-Huntington method?

    a. 1

    b. 2

    c. 3

9. The geometric mean of 7 and 8 is
    a. 7.5.
    b. more than 7.5.
    c. less than 7.5.

10. The relative difference between 7 and 8 is
    a. 1.
    b. 12.5%.
    c. 14.3%.

Word Search

Refer to pages 531 – 532 of your text to obtain the Review Vocabulary. There are 22 hidden vocabulary words/expressions in the word search below. $\lfloor q \rfloor$ and $\lceil q \rceil$ do not appear in the word search. It should be noted that spaces are removed as well as hyphens. *Critical Divisor* appears in the word search.

```
D N O D O H T E M R O S I V I D D R A D N A T S A
E L E U E C N E R E F F I D E V I T A L E R D N M
S E L L D R V T L R N H H L T A S N D O G O R G L
Q A Z X M N N O I T I D N O C A T O U Q H R O G T
N O A G W J E L M N T N E R G Z R P W T V E V L M
P A R S D A T O U Q R E W O L N I E E L R P G A C
P B I G R P O E M N H H E Y U C C M E R R R O L W
K S S M O P R E P R E S E N T A T I V E S H A R E
P O G T J O P O P U L A T I O N P A R A D O X E B
B L T D E R H A M M V C R A E C O E A R N P U T S
R U G O U T D T D T C T O M S R P H D A J X E A T
V T E H P I B K M Q F O N E O I U N J E E J S E E
S E O T P O N D O R E O M F K T L N U H F J G R R
R D M E E N R G M E I E W V A I A A S E F H A E M
R I E M R M E H M T T N J A E C T N T F E C B S E
H F T N Q E I T R K Y P L E J A I S E T R Q E P T
F F R O U N E O N H H J E F O L O H D E S E K O H
U E I T O T P E J C M F E T M D N K Q U O F S C O
Q R C L T P O P S X X S E M A I S L U H N H A G D
R E M I A R E A T O U Q N W G V E M O C M T A H B
T N E M N O I T R O P P A E V I T A T N E T M F L
E C A A P B E N X U S E H P T S Z R A G T E T G W
S E N H I L L H U N T I N G T O N M E T H O D C F
O I E X A E N L Q I C M T O I R G O O R O F R E H
A L A B A M A P A R A D O X L I S C G R D C A N E
```

1. _____   12. _____
2. _____   13. _____
3. _____   14. _____
4. _____   15. _____
5. _____   16. _____
6. _____   17. _____
7. _____   18. _____
8. _____   19. _____
9. _____   20. _____
10. _____   21. _____
11. _____   22. _____

Chapter 15
Game Theory: The Mathematics of Competition

Chapter Objectives

Check off these skills when you feel that you have mastered them.

☐ Apply the minimax technique to a game matrix to determine if a saddlepoint exists.

☐ When a game matrix contains a saddlepoint, list the game's solution by indicating the pure strategies for both row and column players and the playoff.

☐ Interpret the rules of a zero-sum game by listing its payoffs as entries in a game matrix.

☐ From a zero-sum game matrix whose payoffs are listed for the row player, construct a corresponding game matrix whose payoffs are listed for the column player.

☐ If a two-dimensional game matrix has no saddlepoint, write a set of linear probability equations to produce the row player's mixed strategy.

☐ If a two-dimensional game matrix has no saddlepoint, write a set of linear probability equations to produce the column player's mixed strategy.

☐ When given either the row player or the column player's strategy probability, calculate the game's payoff.

☐ State in your own words the minimax theorem.

☐ Apply the principle of dominance to simplify the dimension of a game matrix.

☐ Construct a bimatrix model for an uncomplicated two-person game of partial conflict.

☐ Determine from a bimatrix when a pair of strategies is in equilibrium.

☐ Understand the role of sophisticated (vs. sincere) voting and the true power of a chair in small committee decision-making.

☐ Construct the game tree for a simple truel.

☐ Analyze the game tree of a truel, using backward induction to eliminate branches.

Guided Reading

Introduction

In competitive situations, parties in a conflict frequently have to make decisions which will influence the outcome of their competition. Often, the players are aware of the options – called **strategies** – of their opponent(s), and this knowledge will influence their own choice of strategies. Game theory studies the **rational choice** of strategies, how the players select among their options to optimize the outcome. Some two-person games involve **total conflict,** in which what one player wins the other loses. However, there are also games of **partial conflict,** in which cooperation can often benefit the players.

Section 15.1   Two-Person Total-Conflict Games: Pure Strategies

⚷ Key idea

The simplest games involve two players, each of whom has two strategies. The payoffs to each of the players is best described by a 2×2 **payoff matrix**, in which a positive entry represents a payoff from the column player to the row player, while a negative entry represents a payment from the row player to the column player.

ᏻ Example A

Consider the following payoff matrix.

$$
\begin{array}{c c}
 & A \quad B \\
\begin{array}{c} C \\ D \end{array} &
\begin{bmatrix} 3 & 4 \\ 2 & -5 \end{bmatrix}
\end{array}
$$

a) If the row player chooses C and the column player chooses B, what is the outcome of the game?
b) If the row player chooses C, what is the minimum payoff he can obtain?
c) If the row player chooses D, what is the minimum payoff he can obtain?
d) If the column player chooses A, what is the most she can lose?
e) If the column player chooses B, what is the most she can lose?

Solution

a) The payoff associated with this outcome is the entry in Row C and Column B. The outcome is 4.
b) The minimum payoff is 3
c) An outcome of -5 means that the row player loses 5 to the column player.
d) The outcome is 3. An outcome of 3 means that the column player loses 3 to the row player.
e) The outcome is 4

⚷ Key idea

We see in these examples that the row player can guarantee himself a payoff of at least 3 by playing C, and that the column player can guarantee that she will not lose more than three by playing A. The entry 3 is the minimum of its row, and it is larger than the minimum of the second row, -5. 3 is thus the **maximin,** and choosing C is the row player's **maximin strategy.** Similarly, 3 is the maximum of column A, and it is smaller than 4, which is the maximum of column B. Hence, 3 is the **minimax** of the columns, and if the column player chooses A, then she is playing her **minimax strategy.** When the maximin and minimax coincide, the resulting outcome is called a **saddlepoint.** The saddlepoint is the **value** of the game, because each player can guarantee at least this value by playing his/her maximin and minimax strategies. However, not every game has a saddlepoint. Games which do not will be studied in the next section.

🌀 Example B

Consider a game in which each of the players (John and Jane Luecke) has a coin, and each chooses to put out either a head or a tail. (Note: The players do not flip the coins.) If the coins match, Jane (the row player) wins, while if they do not match, John (column player) wins. The payoffs are as follows.

		John	
		Head	Tail
Jane	Head	2	−4
	Tail	−3	5

a)   What is the row player's maximin?

b)   What is the column player's minimax?

c)   Does this game have a saddlepoint?

Solution

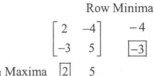

Row Minima

$$\begin{bmatrix} 2 & -4 \\ -3 & 5 \end{bmatrix} \quad \begin{matrix} -4 \\ \boxed{-3} \end{matrix}$$

Column Maxima $\boxed{2}$ 5

a)   −3

b)   2

c)   No. If the maximin is different from the minimax, then there is no saddlepoint.

📝 Question 1

Consider the following payoff matrix.

$$\begin{bmatrix} 3 & 7 \\ 2 & 6 \\ 6 & 9 \end{bmatrix}$$

a)   What is the row player's maximin?

b)   What is the column player's minimax?

c)   Does this game have a saddlepoint?

Answer

a)   6

b)   6

c)   Yes, 6.

Section 15.2   Two-Person Total-Conflict Games: Mixed Strategies

⛏ Key idea

When a game fails to have a saddlepoint, the players can benefit from using mixed **strategies**, rather than **pure strategies**.

⛏ Key idea

The notion of **expected value** is necessary in order to calculate the proper mix of the players' strategies.

🖎 Example C

What is the expected value of a situation in which there are four payoffs, $3, $4, −$2, and $7, which occur with probabilities 0.2, 0.3, 0.45, and 0.05, respectively?

Solution

The expected value is found by multiplying each payoff by its corresponding probability and adding these products. We obtain the following.

$$\$3(0.2)+\$4(0.3)-\$2(0.45)+\$7(0.05)=\$1.25$$

✐ Question 2

What is the expected value of a situation in which there are four payoffs, $2, −$4, $4, and $9, which occur with probabilities 0.25, 0.15, 0.45, and 0.15, respectively?

Answer

$3.05

🖎 Example D

Let's reconsider the game of matching coins, described by the following payoff matrix.

		John Head	John Tail	
		Head	Tail	
Jane	Head	2	−4	q
	Tail	−3	5	$1-q$
		p	$1-p$	

a) Suppose the row player, Jane, mixes her strategy by choosing head with probability q and tails with probability $1-q$. If the column player always chooses heads, what is the row player's expected value?

b) Suppose the row player, Jane, mixes her strategy by choosing head with probability q and tails with probability $1-q$. If the column player always chooses tails, what is the row player's expected value?

c) Find the best value of q, that is, the one which guarantees row player the best possible return. What is the (mixed-strategy) value in this case?

d) Is this game fair?

Solution

a) The expected value is $E_{Head}=2q+(-3)(1-q)=2q-3+3q=-3+5q$.

b) The expected value is $E_{Tail}=-4q+(5)(1-q)=-4q+5-5q=5-9q$.

c) The optimal value of q can be found in this case by setting E_{Head} equal to E_{Tail}, and solving for q

$$E_{Head}=E_{Tail}\Rightarrow-3+5q=5-9q\Rightarrow14q=8$$

$$q=\tfrac{8}{14}=\tfrac{4}{7}\Rightarrow1-q=1-\tfrac{4}{7}=\tfrac{3}{7}$$

Jane's optimal mixed strategy is $(q,1-q)=\left(\tfrac{4}{7},\tfrac{3}{7}\right)$.

To find the value, substitute the q into E_{Head} or E_{Tail}.

The value is $E_{Head}=E_{Tail}=E=5-9\left(\tfrac{4}{7}\right)=\tfrac{35}{7}-\tfrac{36}{7}=-\tfrac{1}{7}$.

d) Since the value of the game is negative, it is unfair to the row player (Jane).

✎ Question 3

Let's reconsider the game of matching coins, described by the following payoff matrix.

		John Head	John Tail	
Jane	Head	2	−4	q
	Tail	−3	5	$1-q$
		p	$1-p$	

a) Suppose the column player, John, mixes his strategy by choosing head with probability p and tails with probability $1-p$.   If the row player always chooses heads, what is the column player's expected value?

b) Suppose the column player, John, mixes his strategy by choosing head with probability p and tails with probability $1-p$.   If the row player always chooses tails, what is the column player's expected value?

c) Find the best value of p, that is, the one which guarantees column player the best possible return.

Answer

a) $-4+6p$

b) $5-8p$

c) $\frac{9}{14}$

☞ Key idea

A game in which the payoff to one player is the negative of the payoff to the other player is called a **zero-sum game**. A zero-sum game can be **non-symmetrical** and yet fair.

ᐺ Example E

Consider a coin-matching game with the following payoff matrix.

		John Head	John Tail
Jane	Head	2	0
	Tail	−1	−3

a) Is the game non-symmetrical?

b) Is it fair?

Solution

a) It is non-symmetrical because the payoffs for the row player are different from those for the column player.

b) It is fair because the value of the game is 0; that payoff, when the row player chooses "head" and the column player chooses "tail," is a saddlepoint.

⌨ Key idea

The **minimax theorem** guarantees that there is a unique game value and an optimal strategy for each player. If this value is positive, then the row player can realize at least this value provided he plays his optimal strategy. Similarly, the column player can assure herself that she will not lose more than this value by playing her optimal strategy. If either one deviates from his or her optimal strategy, then the opponent may obtain a payoff greater than the guaranteed value.

Section 15.3　Partial-Conflict Games

⌨ Key idea

In a game of total conflict, the sum of the payoffs of each outcome is 0, since one player's gain is the other's loss. **Variable-sum games,** on the other hand, are those in which the sum of the payoffs at the different outcomes varies. These are games of partial conflict, because, through cooperation, the players can often achieve outcomes that are more favorable than would be obtained by being pure adversaries.

⌨ Key idea

In many games of partial conflict, it is difficult to assign precise numerical payoffs to the outcomes. However, the preferences of the parties for the various outcomes may be clear. In such a case, the payoffs are **ordinal**, with 4 representing the best outcome, 3 the second best, 2 next, and 1 worst. The payoff matrix now consists of pairs of numbers, the first number representing the row player's payoff, with the second number of the pair being the column player's payoff. Now, both like high numbers.

⌨ Example F

Consider the following matrix

	A	B
C	$(1,3)$	$(2,2)$
D	$(4,1)$	$(3,4)$

a) If the row player chooses D and the column player chooses B, what will the payoffs be to the players?

b) Does either player have a dominant strategy?

Solution

a) The first entry in the outcome $(3,4)$ represents the payoff to the row player, and the second entry, the payoff to the column player. The payoffs will be 3 to the row player and 4 to the column player.

b) The row player gets a better payoff in both cases by choosing strategy D (4 to 1 if the column player selects strategy A, and 3 to 2 if the column player selects strategy B).
The column player gets a more desirable payoff by switching from A to B when the row player selects strategy D; however, she gets a less desirable payoff by making the same switch when the row player selects strategy C. Thus, C is a dominant strategy for the row player. The column player does not have a dominant strategy.

⌨ Key idea

When neither player can benefit by departing unilaterally from a strategy associated with an outcome, the outcome constitutes a **Nash equilibrium**.

ᕙᑊ Example G

Consider the following matrix

	A	B
C	(1,3)	(2,2)
D	(4,1)	(3,4)

a) If this outcome in the matrix is $(1,3)$, does either of the players benefit from defecting?

b) Is there a Nash equilibrium in this matrix?

Solution

a) In the outcome $(1,3)$, the defection from C to D for the row player increases his payoff from 1 to 4. The defection from A to B for the column player, however, produces a payoff decrease. The row player benefits by defecting to D, since he then obtains his best outcome (4), rather than his worst (1).

b) Yes. $(3,4)$ is a Nash equilibrium. Neither player can benefit by changing his or her strategy.

ᑒ Key idea

Prisoners' Dilemma is a game with four possible outcomes. Here, A stands for "arm," and D for "disarm."

There are four possible outcomes:

(D,D): Red and Blue disarm, which is *next best* for both because, while advantageous to each, it also entails certain risks.

(A,A): Red and Blue arm, which is *next worst* for both, because they spend needlessly on arms and are comparatively no better off than at (D, D).

(A,D): Red arms and Blue disarms, which is *best for Red* and *worst for Blue*, because Red gains a big edge over Blue.

(D,A): Red disarms and Blue arms, which is *worst for Red* and *best for Blue*, because Blue gains a big edge over Red.

		Blue A	Blue D
Red	A	(A,A)	(A,D)
	D	(D,A)	(D,D)

This matrix is also used to model other situations.

ᏽ Example H
Consider the following matrix

	A	D
A	(10,10)	(0,20)
D	(20,0)	(1,1)

a)   What is the most favorable outcome for the row player? For the column player?

b)   Does the row player have a dominant strategy in Prisoner's Dilemma?   What about column player?

c)   Would it pay for either player to defect from the outcome $(1,1)$?

d)   Would it pay for either player to defect from the outcome $(10,10)$?

e)   Which outcome is a Nash equilibrium in Prisoners' Dilemma?

Solution

a)   For the row player, it is the outcome where he selects D and the column player selects A. For the column player, it is the reverse. When the row player selects D and the column player A, the row player achieves his maximum payoff: 20.
     When the column player selects D and the row player A, the column player achieves her maximum payoff: 20.

b)   The row player always achieves a better payoff by selecting D rather than A (20 to 10 and 1 to 0). The column player fares similarly with D as the dominant strategy.

c)   No. The payoff to each defector would decrease from 1 to 0.

d)   Yes. The payoff to each defector would increase from 10 to 20.

e)   Neither player can benefit by defecting from the outcome $(1,1)$ because each reduces his or her payoff to 0. Thus, $(1,1)$ is the Nash equilibrium.

⌐ Key idea
Chicken is a game with a payoff matrix such as the following.

	Swerve	Don't swerve
Swerve	(2,2)	(1,4)
Don't swerve	(4,1)	(0,0)

ᏽ Example I
Consider the above matrix.

a)   What is the most favorable outcome for the row player? For the column player?

b)   Does the row player have a dominant strategy? How about the column player?

c)   Would it pay for either player to defect from the outcome $(0,0)$?

Solution
a)   For the row player it is when he doesn't swerve and the column player does. It is the reverse for the column player. Each of these produces a maximum payoff: 4.

b)   When the column player selects "swerve," the row player does better by selecting "not swerve;" however, the opposite is true for the row player's selection of "not swerve." Thus neither player has a dominant strategy.

c)   A defection for each player increases the payoff from 0 to 4. It would pay for either player to defect, since he or she thereby obtains his or her most preferred outcome.

ᏮᎢ Example J
Consider the following matrix.

	Swerve	Don't swerve
Swerve	(2,2)	(4,1)
Don't swerve	(1,4)	(0,0)

Are there Nash equilibria in this game of Chicken?

Solution
$(2,2)$ is a Nash equilibrium. Defecting from outcome $(2,2)$ decreases the row player's payoff from 2 to 1 and the column player's payoff from 2 to 1.

Section 15.4   Larger Games

ᎧᎢ Key idea
If one of three players has a dominant strategy in a $3\times3\times3$ game, we assume this player will choose it and the game can then be reduced to a 3×3 game between the other two players. (If no player has a dominant strategy in a three-person game, it cannot be reduced to a two-person game.)

The 3×3 game is not one of total conflict, so the minimax theorem, guaranteeing players the value in a two-person zero-sum game, is not applicable. Even if the game were zero-sum, the fact that we assume the players can only rank outcomes, but not assign numerical values to them, prevents their calculating optimal mixed strategies in it.

The problem in finding a solution to the 3×3 game is not a lack of Nash equilibria. So the question becomes which, if any, are likely to be selected by the players. Is one more appealing than the others?

Yes, but it requires extending the idea of dominance to its successive application in different stages of play.

ᎧᎢ Key idea
In a small group voting situation (such as a committee of three), **sophisticated voting** can lead to Nash equilibria with surprising results. An example is the status quo paradox. In this situation, supporting the apparently favored outcome actually hurts.

ᎧᎢ Key idea
The analysis of a **"truel"** (three-person duel) is very different when the players move sequentially, rather than simultaneously.

ᏮᎢ Example K
Consider a sequential truel in which three perfect marksmen with one bullet each may fire at each other, each with the goal of remaining alive while eliminating the others. If the players act simultaneously, each has a 25% chance of survival. If they act sequentially, each will choose not to shoot, and all will survive.

Solution
If the players are A, B, C, taking turns in that order, A cannot choose to shoot B (or C), because then C (or B) will shoot him next. A must pass. Similarly with B, and then C; none can risk taking a shot. Thus, all survive.

⌐ Key idea

Sequential truels may be analyzed through the use of a **game tree**, examining it from the bottom up through **backward induction**.

⌐ Key idea

The **theory of moves (TOM)** introduces a dynamic element into the analysis of game strategy. It is assumed that play begins in an initial state, from which the players, thinking ahead, may make subsequent moves and countermoves. Backward induction is the essential reasoning tool the players should use to find optimal strategies.

Section 15.5   Using Game Theory

⌐ Key idea

Game theory provides a framework for understanding the rationale behind conflict in our political and cultural world. An example is the confrontation over the budget between the Democrat President Bill Clinton and the Republican Congress that resulted in a shutdown of part of the federal government on two occasions, between November 1995 and January 1996. Government workers were frustrated in not being able to work, and citizens were hurt and inconvenienced by the shutdown.

Homework Help

Exercises 1 – 5
Carefully read Section 15.1 before responding to these exercises.   Pay special attention to the example in Table 15.2. Note: The payoff matrix in Exercise 5 is

$$\begin{bmatrix} -10 & -17 & -30 \\ -15 & -15 & -25 \\ -20 & -20 & -20 \end{bmatrix}.$$

Exercises 6 – 19
Carefully read Sections 15.1 and 15.2 before responding to these exercises.

Exercises 20 – 25
Carefully read Section 15.3 before responding to these exercises.

Exercises 26 – 37
Carefully read Section 15.4 before responding to these exercises.

Do You Know the Terms?

Cut out the following 34 flashcards to test yourself on Review Vocabulary. You can also find these flashcards at http://www.whfreeman.com/fapp7e.

Chapter 15 Game Theory **Backward induction**	Chapter 15 Game Theory **Chicken**
Chapter 15 Game Theory **Condorcet winner**	Chapter 15 Game Theory **Constant-sum game**
Chapter 15 Game Theory **Dominant strategy**	Chapter 15 Game Theory **Dominated strategy**
Chapter 15 Game Theory **Expected value E**	Chapter 15 Game Theory **Fair game**

A two-person variable-sum symmetric game in which each player has two strategies: to swerve to avoid a collision, or not to swerve and cause a collision if the opponent has not swerved. Neither player has a dominant strategy; the compromise outcome, in which both players swerve, is not a Nash equilibrium, but the two outcomes in which one player swerves and the other does not are Nash equilibria.

A reasoning process in which players, working backward from the last possible moves in a game, anticipate each other's rational choices.

A game in which the sum of payoffs to the players at each outcome is a constant, which can be converted to a zero-sum game by an appropriate change in the payoffs to the players that does not alter the strategic nature of the game.

A candidate that defeats all others in separate pairwise contest.

A strategy that is sometimes worse and never better for a player than some other strategy, whatever strategies the other players choose.

A strategy that is sometimes better and never worse for a player than every other strategy, whatever strategies the other players choose.

A zero-sum game is fair when the (expected) value of the game, obtained by using optimal strategies (pure or mixed), is zero.

If each of the n payoffs, $s_1, s_2, ..., s_n$ occurs with respective probabilities $p_1, p_2, ..., p_n$, then
$$E = p_1 s_1 + p_2 s_2 + ... + p_n s_n$$
where $p_1 + p_2 + ... + p_n = 1$ and $p_i \geq 0$ $(i = 1, 2, ..., n)$.

Chapter 15 Game Theory **Game tree**	Chapter 15 Game Theory **Maximin**
Chapter 15 Game Theory **Maximin strategy**	Chapter 15 Game Theory **Minimax**
Chapter 15 Game Theory **Minimax strategy**	Chapter 15 Game Theory **Minimax theorem**
Chapter 15 Game Theory **Mixed strategy**	Chapter 15 Game Theory **Nash equilibrium**

In a two-person zero-sum game, the largest of the minimum payoffs in each row of a payoff matrix.	A symbolic tree, based on the rules of play in a game, in which the vertices, or nodes, of the tree represent choice points, and the branches represent alternative courses of action that the players can select.
In a two-person zero-sum game, the smallest of the maximum payoffs in each column of a payoff matrix.	In a two-person zero-sum game, the pure strategy of the row player corresponding to the maximin in a payoff matrix.
The fundamental theorem for two-person constant-sum games, stating that there always exist optimal pure or mixed strategies that enable the two players to guarantee the value of the game.	In a two-person zero-sum game, the pure strategy of the column player corresponding to the minimax in a payoff matrix.
Strategies associated with an outcome such that no player can benefit by choosing a different strategy, given that the other players do not depart from their strategies.	A strategy that involves the random choice of pure strategies, according to particular probabilities. A mixed strategy of a player is optimal if it guarantees the value of the game.

Chapter 15 Game Theory **Nonsymmetrical game**	Chapter 15 Game Theory **Ordinal game**
Chapter 15 Game Theory **Partial-conflict game**	Chapter 15 Game Theory **Payoff matrix**
Chapter 15 Game Theory **Plurality procedure**	Chapter 15 Game Theory **Prisoners' Dilemma**
Chapter 15 Game Theory **Pure strategy**	Chapter 15 Game Theory **Rational choice**

A game in which the players rank the outcomes from best to worst.	A two-person constant-sum game in which the row player's gains are different from the column player's gains, except when there is a tie.
A rectangular array of numbers. In a two-person game, the rows and columns correspond to the strategies of the two players, and the numerical entries give the payoffs to the players when these strategies are selected.	A variable-sum game in which both players can benefit by cooperation but may have strong incentives not to cooperate.
A two-person variable-sum symmetric game in which each player has two strategies, cooperate or defect. Cooperate dominates defect for both players, even though the mutual-defection outcome, which is the unique Nash equilibrium in the game, is worse for both players than the	A voting procedure in which the alternative with the most votes wins.
A choice that leads to a preferred outcome.	A course of action a player can choose in a game that does not involve randomized choices.

Chapter 15
Game Theory

Saddlepoint

Chapter 15
Game Theory

Sincere voting

Chapter 15
Game Theory

Sophisticated voting

Chapter 15
Game Theory

Status-quo paradox

Chapter 15
Game Theory

Strategy

Chapter 15
Game Theory

Theory of moves (TOM)

Chapter 15
Game Theory

Total-conflict game

Chapter 15
Game Theory

Value

Voting for one's most-preferred alternative in a situation.	In a two-person constant-sum game, the payoff that results when the maximin and the minimax are the same, which is the value of the game. The saddlepoint has the shape of a saddle-shaped surface and is also a Nash equilibrium.
The status quo is defeated by another alternative, even if there is no Condorcet winner, when voters are sophisticated.	Involves the successive elimination of dominated strategies by voters.
A dynamic theory that describes optimal choices in strategic-form games in which players, thinking ahead, can make moves and countermoves.	One of the courses of action a player can choose in a game; strategies are mixed or pure, depending on whether they are selected in a randomized fashion (mixed) or not (pure).
The best outcome that both players can guarantee in a two-person zero-sum game. If there is a saddlepoint, this is the value. Otherwise, it is the expected payoff resulting when the players choose their optimal mixed strategies.	A zero-sum or constant-sum game, in which what one player wins the other player loses.

Chapter 15
Game Theory

Variable-sum game

Chapter 15
Game Theory

Zero-sum game

A constant-sum game in which the payoff to one player is the negative of the payoff to the other player, so the sum of the payoffs to the players at each outcome is zero.

A game in which the sum of the payoffs to the players at the different outcomes varies.

Practice Quiz

1. In the following two-person zero-sum game, the payoffs represent gains to Row Player I and losses to Column Player II.

$$\begin{bmatrix} 3 & 6 \\ 4 & 8 \end{bmatrix}$$

   Which statement is true?

   a. The game has no saddlepoint.

   b. The game has a saddlepoint of value 4.

   c. The game has a saddlepoint of value 6.

2. In the following two-person zero-sum game, the payoffs represent gains to Row Player I and losses to Column Player II.

$$\begin{bmatrix} 4 & 7 & 1 \\ 3 & 9 & 5 \\ 8 & 2 & 6 \end{bmatrix}$$

   What is the maximin strategy for Player I?

   a. Play the first row.

   b. Play the second row.

   c. Play the third row.

3. In the following two-person zero-sum game, the payoffs represent gains to Row Player I and losses to Column Player II.

$$\begin{bmatrix} 4 & 7 & 1 \\ 3 & 9 & 5 \\ 8 & 2 & 6 \end{bmatrix}$$

   What is the minimax strategy for Player II?

   a. Play the first column.

   b. Play the second column.

   c. Play the third column.

4. In a two-person zero-sum game, suppose the first player chooses the second row as the maximin strategy, and the second player chooses the third column as the minimax strategy. Based on this information, which of the following statements is true?

   a. The game definitely has a saddlepoint.

   b. If the game has a saddlepoint, it must be in the second row.

   c. The game definitely does not have a saddlepoint.

5. In the game of matching pennies, Player I wins a penny if the coins match; Player II wins if the coins do not match. Given this information, it can be concluded that the two-by-two matrix which represents this game

   a. has all entries the same.

   b. has entries which sum to zero.

   c. has two 0s and two 1s.

6. In the following game of batter-versus-pitcher in baseball, the batter's batting averages are given in the game matrix.

		Pitcher	
		Fastball	Fastball
Batter	Fastball	0.300	0.200
	Curveball	0.100	0.400

What is the pitcher's optimal strategy?

a. Throw more fastballs than curveballs.

b. Throw more curveballs than fastballs.

c. Throw equal proportions of fastballs and curveballs.

7. In the following game of batter-versus-pitcher in baseball, the batter's batting averages are given in the game matrix.

		Pitcher	
		Fastball	Fastball
Batter	Fastball	0.300	0.200
	Curveball	0.100	0.400

What is the batter's optimal strategy?

a. Anticipate more fastballs than curveballs.

b. Anticipate more curveballs than fastballs.

c. Anticipate equal proportions of fastballs and curveballs.

8. Consider the following partial-conflict game, played in a non-cooperative manner.

		Player II	
		Choice A	Choice B
Player I	Choice A	(3,3)	(4,1)
	Choice B	(1,4)	(2,2)

What outcomes constitute a Nash equilibrium?

a. Only when both players select Choice A.

b. Only when both players select Choice A or both select Choice B.

c. Only when one player selects Choice A and the other selects Choice B.

9. True or False: Sequential and simultaneous trials result in different outcomes.

a. True

b. False

10. True or False: A deception strategy can help deal with the status quo paradox.

a. True

b. False

Word Search

Refer to pages 580 – 581 of your text to obtain the Review Vocabulary. There are 34 hidden vocabulary words/expressions in the word search below. This represents all vocabulary words. It should be noted that spaces and hyphens, are removed as well as apostrophes. Also, the abbreviations do not appear in the word search. The backside of this page has additional space for the words/expressions that you find.

```
N M B Y N I N C C O N S T A N T S U M G A M E Y Y
E A R P I R T R H R Y G E T A R T S A A S M I G F
X X I U E X P E C T E D V A L U E C H W A S E E V
C I P R I S O N E R S D I L E M M A G G Z T M T A
N M L E P O Z E R O S U M G A M E K T D A A A A R
O I G S L P T F F X E Z E T E E O C S R X T D R I
N N G T U H P A Y O F F M A T R I X T E R U A T A
S L E R R I G M E O A T E X O L H S I N A S N S B
Y Y F A A S Y T A M S P A M F Q D S P N L Q U N L
M T G T L T N D N R A M T N V E D M T I N U E I E
M E N E I I E N I I I G O B T Y M V J W M O M M S
E M A G T C I L F N O C L A I T R A P T I P I I U
T A S Y Y A N S I S L P N A E F O L O E N A X X M
R G H A P T R M E A P I E K N Z E U C C I R E A G
I R E Q R E G T T V M P E L T I I E P R M A D M A
C I Q H O D E O S O O G L H D G D K X O A D S E M
A A U N C V T B D T L M E H N D E R E D X O T E E
L F I F E O D F E M N T F D T M A N O N S X R M T
G L L F D T Q J S N L A E O G H E S N O T E A M R
A I I K U I A G G R R I N S Y G E E E C R G T R E
M X B D R N O I T C U D N I D R A W K C A B E N E
E T R O E G F F M X N I G K M P O V C B T P G G E
E C I O H C L A N O I T A R H O S E I I E L Y E R
C G U S I N C E R E V O T I N G D G H O G R E Y D
T S M I N I M A X T H E O R E M Y S C T Y R I G M
```

1. _____

2. _____

3. _____

4. _____

5. _____

6. _____

7. _____

8. _____

9. _____

10. _____

11. _____

12. _____

13. _____

14. _____

15. _____

16. _____

17. _____

18. _____

19. _____

20. _____

21. _____

22. _____

23. _____

24. _____

25. _____

26. _____

27. _____

28. _____

29. _____

30. _____

31. _____

32. _____

33. _____

34. _____

Chapter 16
Identification Numbers

Chapter Objectives

Check off these skills when you feel that you have mastered them.

☐ Understand the purpose of a check digit and be able to determine one for various schemes.

☐ Given an identification number and the scheme used to determine it, be able to decide if the number is a valid number for that scheme.

☐ Given an identification number and the scheme, use it to decipher the information such as birth date and sex.

☐ Be able to convert a given ZIP code to its corresponding bar code, and vice versa.

☐ Be able to convert a given UPC number to its corresponding bar code.

Guided Reading

Introduction

⌐ **Key idea**

Almost everything we encounter in daily life – consumer goods, credit cards, financial records, people, organizations, mail – is somehow identified or classified by a numeric or alphanumeric code. Of course, this code must unambiguously identify the individual or object it names. But since humans and machines are fallible, the system used for creating the code must be designed to minimize errors. Also, since errors will certainly occur, the system should include a mechanism for detecting and, if possible, correcting the most common errors.

Section 16.1   Check Digits

⌐ **Key idea**

Many frequently used types of error-detecting code for identification numbers include an extra digit (usually the last digit) called a **check digit**. The check digit can be compared to the rest of the number to check for validity.

⌐ **Key idea**

For a postal service money order, the check digit (the last digit) is the remainder you get when you divide the sum of the other digits by 9.

ᘒᓀ Example A

If the ID number on a postal money order is 321556738X, what is the value of X?

Solution

The other digits add up to 40; divide by 9, the remainder is 4. Thus, X = 4.

⌗ᓂ Key idea

Some mail (UPS) and car rental services use as an extra check digit as the remainder when you divide the number by 7.

ᘒᓀ Example B

If the ID number of a FedEx package is 321556738X, what is the value of X?

Solution

If you divide 321556738 by 7, you get 45936676 with a remainder of 6. Thus, X = 6.

✎ Question 1

If the ID number of a FedEx package is 3213213213X, what is the value of X?

Answer

0

⌗ᓂ Key idea

The *Universal Product Code* (*UPC*) is a twelve-digit code, $a_1a_2a_3a_4a_5a_6a_7a_8a_9a_{10}a_{11}a_{12}$, including a check digit at the end. By adding the digits and multiplying by their weight (alternately 1 for even positions, 3 for odd positions), the sum must be a number ending in 0.

ᘒᓀ Example C

If the UPC code for a product is 43276598731X, what is the value of X?

Solution

Since $3 \cdot 4 + 3 + 3 \cdot 2 + 7 + 3 \cdot 6 + 5 + 3 \cdot 9 + 8 + 3 \cdot 7 + 3 + 3 \cdot 1 = 113$, the last digit (check digit) must be 7 in order to have a sum that ends in 0.

✎ Question 2

If the UPC code for a product is 12345678901X, what is the value of X?

Answer

2

⌗ᓂ Key idea

The **Codabar** system is a variation of UPC using a similar sum with weights 2 (odd positions) and 1 (even positions). To this sum, you add the number of digits in odd positions that exceed 4; the resulting number must end in 0 to be a valid Codabar code.

ᕫᐧ Example D
Is 4327 6598 2341 7112 a valid credit card number using Codabar?

Solution
The numbers in the odd positions are 4_2_ 6_9_ 2_4_ 7_1_. The numbers in the even positions are _3_7 _5_8 _3_1 _1_2 (the last position is the check digit). The numbers that exceed 4 in odd positions are _ _ _ _ 6_9_ _ _ _ _ 7 _ _ _.
The Codabar algorithm yields the following.

$$(4+2+6+9+2+4+7+1) \times 2 + 3 + (3+7+5+8+3+1+1+2) = 103$$

Since this number does not end in zero (103), the credit card is not a valid number using the Codabar algorithm.

✐ Question 3
What is the missing digit of the fictitious credit card number 1234 5678 X234 5678 using Codabar?

Answer
2

⛊ Key idea
Another important and effective error-detecting code is the **International Standard Book Number** (**ISBN**). It is a 10-digit code, $a_1 a_2 a_3 a_4 a_5 a_6 a_7 a_8 a_9 a_{10}$. It has the following property.

$$10a_1 + 9a_2 + 8a_3 + 7a_4 + 6a_5 + 5a_6 + 4a_7 + 3a_8 + 2a_9 + a_{10} \text{ is evenly divisible by } 11.$$

ᕫᐧ Example E
Explain why the ISBN of this *Guide* is valid.

Solution
The ISBN of this manual is 0-7167-6946-8. Using the algorithm we have the following.
$$10 \cdot 0 + 9 \cdot 7 + 8 \cdot 1 + 7 \cdot 6 + 6 \cdot 7 + 5 \cdot 6 + 4 \cdot 9 + 3 \cdot 4 + 2 \cdot 6 + 8 = 253$$
Since 253 is divisible by 11 ($11 \cdot 23 = 253$), this is a valid ISBN.

Section 16.2   The ZIP Code

⛊ Key idea
The **ZIP code** is a U. S. Postal Service ID that encodes geographical information about each post office. The first digit in a ZIP code represents one of the ten regions, from east to west, numbered 0–9. Each state is divided into a variable number of smaller geographical areas. The second two digits represent the central mail-distribution point in this area.

⛊ Key idea
ZIP + 4 code is a further refinement of the ZIP code, capable of identifying small groups of mailboxes, like a floor of a building, within a given postal zone.

Section 16.3   Bar Codes

⛊ Key idea
Bar codes use light spaces and dark bars to represent a two-symbol binary code that is easily scanned optically and decoded by a computer.

🔑 Key idea

ZIP code bar codes use the **postnet code**. Each digit is represented by a group of five dark bars, two long and three short. At the beginning and end there are two long bars, which are called guard bars. The following are the bar patterns.

ıııll 1   ıllıı 6

ıılıl 2   lıııl 7

ıılⅼı 3   lılıı 8

ılııl 4   lılⅼı 9

ılⅼlı 5   llııı 0

👌 Example F

What would the bar code look like for a ZIP of 53207?

Solution

With guard bars at the beginning and end, we have the following.

5   3   2   0   7

🔑 Key idea

A ZIP + 4 number is made up of 9 digits and one check digit. Including a check digit at the end, the sum of the ZIP + 4 digits must be a number ending in 0.

✏️ Question 4

Render a drawing of what the bar code would look like for the following postcard, including the check digit for the ZIP + 4 number. What would the check digit be for the postcard?

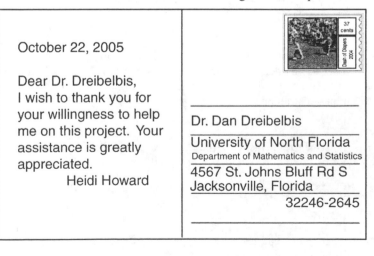

October 22, 2005

Dear Dr. Dreibelbis,
I wish to thank you for
your willingness to help
me on this project. Your
assistance is greatly
appreciated.
　　　　　Heidi Howard

Dr. Dan Dreibelbis
University of North Florida
Department of Mathematics and Statistics
4567 St. Johns Bluff Rd S
Jacksonville, Florida
　　　　　32246-2645

Answer

6

Key idea

The **UPC (Universal Product Code)** is a familiar sight on labels for retail products. Digits are represented by sequences of light and dark bars, where adjacent dark bars blend together to make bars of different widths. In this way, seven bar spaces ("modules") produce two light and two dark bars for each digit. There are different binary coding patterns for manufacturer numbers and product numbers. Refer to Table 16.1, page 608 of your textbook.

Example G

How is the digit 7 represented in the UPC bar code?

Solution

For manufacturer code, 7 is equivalent to 0111011. The bar pattern would look something like the following.

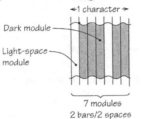

For product code, 7 is equivalent to 1000100. The bar pattern would look something like the following.

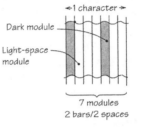

Section 16.4   Encoding Personal Data

Key idea

Personal data such as your name, birthdate, or sex can be encoded different ways. One way to encode a surname (last name) is the **Soundex Coding System**. The procedure is as follows.

1. Delete all occurrences of h and w.
2. Assign numbers to the remaining letters as follows:

$$a, e, i, o, u, y \rightarrow 0$$
$$b, f, p, v \rightarrow 1$$
$$c, g, j, k, q, s, x, z \rightarrow 2$$
$$d, t \rightarrow 3$$
$$l \rightarrow 4$$
$$m, n \rightarrow 5$$
$$r \rightarrow 6$$

3. If two or more letters with the same numeric value are adjacent, omit all but the first.
4. Delete the first character of the original name if still present.
5. Delete all occurrences of $a, e, i, o, u,$ and y.
6. Retain only the first three digits corresponding to the remaining letters; append trailing 0's if fewer than three letters remain; precede the three digits obtained in step 6 with the first letter of the surname.

Example H

Encode the surname *Howard* using the Soundex Coding System.

Solution

	Step 1	Step 2	Step 3	Step 4	Step 5	Step 6
Howard	$\rightarrow$ oard	$\rightarrow$ oard	$\rightarrow$ ord	$\rightarrow$ ord	$\rightarrow$ rd	$\rightarrow$ H-630
		0063	063	063	63	

The surname Howard is encoded as H-630.

✐ Question 5

How would the surname *Hochwald* be encoded using the Soundex Coding System?

Answer

H-243

⚷ Key idea

Florida and Illinois driver's licenses use an encoding system that considers both the date of birth and the sex of the resident.

- Florida: A five-digit number contains the information for the resident. The first two digits are the birth year (without the 19), followed by a dash. Thus, 1966 is represented as 66. Each day of the year is assigned a three-digit number with 001 representing January 1. Each month is assigned 40 days. Thus, April 3 is 123. Added to this number is 500 if the resident is female. Thus, a female born on April 3, 1966 is given the number 66-623.

- Illinois: A five-digit number contains the information for the resident. The first two digits (separated by a dash) are the birth year (without the 19). Thus, 1966 is represented as 6-6. Each day of the year is assigned a three-digit number with 001 representing January 1. Each month is assigned 31 days. Thus, April 3 is 096. Added to this number is 600 if the resident is female. Thus, a female born on April 3, 1966 is given the number 6-6696.

Ꮽ Example I

What would be the five-digit number assigned to a male driver in the state of Florida born on October 20, 1977?

Solution

77-920

✐ Question 6

5-3640 identifies a Florida driver of which sex and birth date?

Answer

Female born on April 20, 1953.

✐ Question 7

5-3640 identifies an Illinois driver of which sex and birth date?

Answer

Female born on February 9, 1953.

Homework Help

Exercises 1 – 46
Carefully read Section 16.1 before responding to these exercises. These exercises involve examining check digits. Make sure you understand clearly what method in calculating check digits is used before answering the exercise.

Exercises 47 – 53
Carefully read Sections 16.2 and 16.3 before responding to these exercises. Have the bar codes in front of you and remember that there is a guard bar at the beginning and end of each bar code. Each grouping of five bars is made up of three short and two long bars.

Exercises 54 – 56
Carefully read Section 16.3 before responding to these exercises.

Exercises 57 – 61
Carefully read Spotlight 16.4 (page 609) before responding to these exercises.

Exercises 62 – 63
Carefully read Section 16.4 before responding to these exercises. The Soundex Coding System is a six-step process as explained on page 610 of the text.

Exercises 64 – 71
Carefully read Section 16.4 before responding to these exercises. These exercises involve the sex and birthdate on driver's licenses. The encoding scheme is either given in the problem or is explained in the last part of Section 16.4.

Exercises 72 – 75
Carefully read Section 16.4 before responding to these exercises. Answers may vary as you think about possible explanations.

Do You Know the Terms?

Cut out the following 14 flashcards to test yourself on Review Vocabulary. You can also find these flashcards at http://www.whfreeman.com/fapp7e.

Chapter 16 **Identification Numbers** **Bar code**	**Chapter 16** **Identification Numbers** **Binary code**
Chapter 16 **Identification Numbers** **Check digit**	**Chapter 16** **Identification Numbers** **Codabar**
Chapter 16 **Identification Numbers** **Decoding**	**Chapter 16** **Identification Numbers** **Encoding**
Chapter 16 **Identification Numbers** **Error-detecting code**	**Chapter 16** **Identification Numbers** **International Standard Book Number (ISBN)**

A coding scheme that uses two symbols, usually 0 and 1.

A code that employs bars and spaces to represent information.

An error-detection method used by all major creditcard companies, many libraries, blood banks, and others.

A digit included in an identification number for the purpose of error detection.

Translating data into code.

Translating code into data.

A 10-digit identification number used on books throughout the world that contains a check digit for error detection.

A code in which certain types of errors can be detected.

Chapter 16
Identification Numbers

Postnet code

Chapter 16
Identification Numbers

Soundex Coding System

Chapter 16
Identification Numbers

Universal Product Code (UPC)

Chapter 16
Identification Numbers

Weights

Chapter 16
Identification Numbers

ZIP code

Chapter 16
Identification Numbers

ZIP + 4 code

An encoding scheme for surnames based on sound.	The bar code used by the U.S. Postal Service for ZIP codes.
Numbers used in the calculation of check digits.	A bar code and identification number that are used on most retail items. It detects 100% of all single-digit errors and most other types of errors.
The nine-digit code used by the U.S. Postal Service to refine ZIP codes into smaller units.	A five-digit code used by the U.S. Postal Service to divide the country into geographic units to speed sorting of the mail.

Practice Quiz

1. Suppose a U. S. Postal Service money order is numbered 632930421#, where the last digit is obliterated. What is the missing digit?

   a. 3

   b. 6

   c. 0

2. Suppose an American Express Travelers Cheque is numbered #293019225, where the first digit is obliterated. What is the missing digit?

   a. 3

   b. 4

   c. 6

3. Is the number 3281924 a legitimate Avis rental car number?

   a. Yes

   b. No, but if the final digit is changed to a 2, the resulting number 3281922 is legitimate.

   c. No, but if the final digit is changed to a 3, the resulting number 3281923 is legitimate.

4. Is the number 3234580005 a legitimate airline ticket number (assume number of digits is acceptable)?

   a. Yes

   b. No, but if the final digit is changed to a 4, the resulting number 3234580004 is legitimate.

   c. No, but if the final digit is changed to a 2, the resulting number 3234580002 is legitimate.

5. Determine the check digit that should be appended to the UPC code 0-10010-34500.

   a. 2

   b. 4

   c. 8

6. Determine the check digit that should be appended to the bank identification number 015 000 64.

   a. 2

   b. 5

   c. 8

7. Determine the check digit that should be appended to the Codabar number 312580016535003.

   a. 1

   b. 3

   c. 7

8. Suppose the ISBN 0-1750-3549-0 is incorrectly reported as 0-1750-3540-0. Which of the following statements is true?

   a. The check digit will detect the error, but cannot correct it.

   b. The check digit will detect and correct the error.

   c. The check digit cannot detect the error.

9. How would the surname *Lee* be encoded using the Soundex Coding System?

    a. L-000

    b. L-040

    c. L-400

10. Suppose that a Postnet code is incorrectly reported. You know that only one of the digits is incorrectly reported. Which of these statements is true?

    a. If you know which digit is incorrect, you can always correct a single error in a Postnet code.

    b. If you know which digit is incorrect, you can sometimes but not always correct a single error in a Postnet code.

    c. Even if you know which digit is incorrect, you can never correct a single error in a Postnet code.

Word Search

Refer to pages 611 – 612 of your text to obtain the Review Vocabulary. There are 12 hidden vocabulary words/expressions in the word search below. It should be noted that spaces are removed as well as hyphens. *Zip + 4 code* and *International Standard Book Number* (*ISBN*) do not appear. Also, the abbreviations do not appear in the word search.

```
S I H A E E M O A E N Q S I G W H A K G M L E S A
M N T P S L E D O E P E N N D L X U C G S K E E A
E T F O A C T D R L J M A H H Q Z Q E S M G B N I
A C I P L A K O C N G X H O Q S X E E E N A M E I
E H H D N I D W H S I T Q Y L G N I D O C N E C E
M E T S Y S G N I D O C X E D N U O S N R A F A E
I C F R O X N S S S A E H O E N C I S I A A W E N
N K J A S E D O C G N I T C E T E D R O R R E O F
R D H E H B A E H R A R H M C Q T N W J E J V R C
L I P S T W S E R K R N E U F H B W T A M I E R N
O G E J S E E R E D S E D T E L Y K I G C F E M S
E I C N E I N P O D J O T I N U N A O H J L C O F
S T R X L C O S E D R A F R N C O N S A S K M F Y
C N A P W E T Z I P C O D E E I O O O K M N I S F
E B S J J H E B L O A H N R O G U D O O E L U N H
B Z R L G T R A B S H T E D O C Y R A N I B S T P
E D N I T N S Y A T Q A J X E M O K T B X Y N O N
A U E K A R I R R N Z T E D J R E O G T A Y D N E
I W E E E X S D C E S E F F R L C J R F F R I V E
F A G V O E L R O T P P C T S H C F D G A M R F E
P I I E A M O G D C K E T R C E O A T A E C E A T
S N S K L S E H O E I B R D O M O A R N O G H Q
U I E P M H V M H D T D E S W F A B G C B G G L G
E R R J L I U E T E M N P E I G N T A Y N M A J S
I N N T H L A G O A A N A Z M S E R O W E O T C Q
```

1. _____   7. _____

2. _____   8. _____

3. _____   9. _____

4. _____   10. _____

5. _____   11. _____

6. _____   12. _____

Chapter 17
Information Science

Chapter Objectives

Check off these skills when you feel that you have mastered them.

- [ ] Know what a binary code is.

- [ ] Use the diagram method to determine or verify a code word, given the message.

- [ ] Use the diagram method to decode a received message.

- [ ] Be able to compute check digits for code words given the parity-check sums for the code.

- [ ] Be able to determine the distance between two n-tuples of 0's and 1's.

- [ ] Be able to determine the weight of a code word and the minimum weight of the nonzero code words in a code (for binary codes, the minimum weight is the same as the minimum distance between code words).

- [ ] Know what nearest-neighbor decoding is and be able to use it for decoding messages received in the Hamming code of Table 17.2.

- [ ] Be able to encode and decode messages that have symbols (such as letters of the alphabet) expressed in binary form.

- [ ] Be able to make observations regarding frequently (or infrequently) occurring letters.

- [ ] Be able to decode using a Huffman code and be able to create a Huffman code, given a table of probabilities.

- [ ] Be able to encode and decode messages using the Caesar and Vigenère ciphers.

- [ ] Be able to add binary strings.

- [ ] Be able to perform calculations using modular arithmetic.

- [ ] Understand how the RSA public key encryption scheme works.

- [ ] Be able to complete a truth table given 2 or 3 statements and the connectives NOT, OR, and AND.

- [ ] Be able to determine if two statements are logically equivalent.

Guided Reading

Introduction

In this chapter we consider some sophisticated techniques to detect and correct errors in digitally transmitted messages. We also take a look at methods that have been developed for the compression of data and for protection of the confidentiality of our messages. The Internet was originally used only by a few organizations as a source of information, but use of the Internet has become far more widespread. Now the Internet is used for shopping, entertainment, games, and much more. Web search efficiency is of importance with such wide spread use.

Section 17.1   Binary Codes

⌖ Key idea

Most computerized data are stored and transmitted as sequences of 0's and 1's. We can store a data string of length 4 in regions I through IV, respectively.

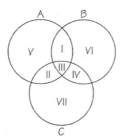

Regions V, VI, and VII have appended digits so that the sum of the regions for each circle has **even parity**. The encoded messages are called **code words**. This scheme is helpful to detect and even correct errors.

ᘓ Example A

Store the data string 1011 along with its appended digits correctly in a three-circle diagram. What is the code word?

Solution

Original data string

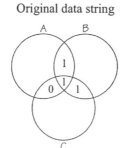

Coded data string

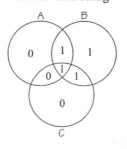

The code word is 1011010.

✎ Question 1

If a code word was received as 0100011, what region (if any) would be changed to decode the message using the diagram method?

Answer

Region III

Section 17.2   Encoding with Parity-Check Sums

⌁→ Key idea

If a string of digits $a_1\ a_2\ a_3\ a_4$ is a message, then we can create a code word in a **binary linear code** by adding check digits that are the **parity-check sums** $a_1 + a_2 + a_3$, $a_1 + a_3 + a_4$, and $a_2 + a_3 + a_4$.

ᕬ Example B

What is the code word for the message 1100?

Solution

$a_1 + a_2 + a_3 = 2$, even parity; so the first check digit is 0. Both $a_1 + a_3 + a_4$ and $a_2 + a_3 + a_4 = 1$, odd parity; so the last two check digits are 1. The code word is therefore, 1100011.

⌁→ Key idea

Since there may be errors in the coding process or transmission, we use the **nearest-neighbor decoding method** to find the code word with the shortest distance (number of positions in which the strings differ) from the received message. Use Table 17.1 on page 620.

ᕬ Example C

Using the nearest-neighbor method, decode these two code words.
a)   1110110
b)   0110010

Solution

a)   The distance between the strings is 1, the error is in the sixth digit. 1110110 should be 1110100.
b)   The distance is 0, there is no error. The message is a valid code word.

⌁→ Key idea

The weight of a binary code is the minimum number of 1's that occur among all nonzero code words of that code.

⌁→ Key idea

In a **variable-length code**, we can use **data compression** to make the shortest code words correspond to the most frequently occurring strings. Morse code is an example.

⌁→ Key idea

In encoding a sequence of letters with a binary code (or decoding the binary code), we need an assignment of letters and code.

ᕬ Example D

Use $A \to 0$, $B \to 10$, $C \to 110$, $D \to 1110$, $E \to 11110$, $F \to 111110$, $G \to 1111111$ to decode the following.

$$0111011110111111111111011011100$$

Solution

Noticing the location of the zeros, 0111011110111111111111011011100 can be written as 0, 1110, 11110, 1111111,111110, 110, 1110, 0. Thus we have, *ADEGFCDA*.

🔑 Key idea

Huffman coding assigns to letters of the alphabet strings of variable size depending on the frequency a letter occurs or an assigned probability. To do this one needs to create a **code tree** (binary tree) performing the following steps.

- List the letters in increasing order top to bottom in terms of their probabilities. Note: The sum of the probabilities will be 1.
- The two letters that have the lowest probability of occurring get grouped together. Their probabilities are summed and the letter that has the smallest probability of occurring appears on the left. Rearrange the letters (two are grouped together), if necessary, in terms of their probabilities of occurring. From here on out a group of letters can consist of more than one letter.
- Continue this process with the new list until all letters appear in a string and the probability is 1.
- The letters/combination of letters are now spread out into a tree, working backwards. Figure 17.7 on page 633 demonstrates this with *EC* having probability 0.425 placed with a 0 on the top branch and *FDBA* having probability 0.575 placed with a 1 on the bottom branch.
- Read the variable string code for the letters right to left.

✏️ Question 2

Use a Huffman tree code to assign a binary code given the following.

A	B	C	D	E	F	G
0.105	0.235	0.100	0.115	0.195	0.107	0.143

Answer

$A = 001, B = 10, C = 000, D = 011, E = 111, F = 010, G = 110.$

Section 17.3   Cryptography

🔑 Key idea

Encryption of stored and transmitted data protects its security. For example, passwords are stored in encrypted form in computers.

🔑 Key idea

The so-called **Caesar cipher** is a cryptosystem that assigns a letter of the alphabet to another letter of the alphabet by shifting each letter of the alphabet by a fixed algorithm. Since there is a limited number of letters of the alphabet, there is a limited number of possible shifts.

🔑 Key idea

Modular arithmetic can be used in encrypting information. The notation $a \bmod n$ is read as a modulo n. a and n are both positive integers. $a \bmod n$ is the remainder when a is divided by n.

🖊 Example E

Calculate the following.

a) $41 \bmod 26$ c) $7 \bmod 12$

b) $112 \bmod 11$ d) $8 \bmod 8$

Solution

a) $41 \bmod 26 = 15$ because $41 = 1 \cdot 26 + 15.$ c) $7 \bmod 12 = 7$ because $7 = 0 \cdot 12 + 7.$

b) $112 \bmod 11 = 2$ because $112 = 10 \cdot 11 + 2.$ d) $8 \bmod 8 = 0$ because $8 = 1 \cdot 8 + 0.$

☞ Key idea

The **Vigenère cipher** requires a key word that will first shift each letter of the word to be encoded. Each letter of the word to be encoded is identified by a position $0 - 25$ (A is located in position 0, not 1). The code word then shifts each letter and that result is evaluated modulo 26.

Letter	A	B	C	D	E	F	G	H	I	J	K	L	M
Location	0	1	2	3	4	5	6	7	8	9	10	11	12
Letter	N	O	P	Q	R	S	T	U	V	W	X	Y	Z
Location	13	14	15	16	17	18	19	20	21	22	23	24	25

ᏊᏒ Example F

Use the Vigenère cipher with the keyword PEELING to encrypt the message I'M NOT FEELING WELL. Note: The apostrophe is not encrypted.

Solution

P is in the position 15; E is in position 4; L is in position 11; I is in position 8, N is in position 13, and G is in position 6.

Original	Location			Encrypted
I	8	15	$(8+15)\bmod 26 = 23\bmod 26 = 23$	X
M	12	4	$(12+4)\bmod 26 = 16\bmod 26 = 16$	Q
N	13	4	$(13+4)\bmod 26 = 17\bmod 26 = 17$	R
O	14	11	$(14+11)\bmod 26 = 25\bmod 26 = 25$	Z
T	19	8	$(19+8)\bmod 26 = 27\bmod 26 = 1$	B
F	5	13	$(5+13)\bmod 26 = 18\bmod 26 = 18$	S
E	4	6	$(4+6)\bmod 26 = 10\bmod 26 = 10$	K
E	4	15	$(4+15)\bmod 26 = 19\bmod 26 = 19$	T
L	11	4	$(11+4)\bmod 26 = 15\bmod 26 = 15$	P
I	8	4	$(8+4)\bmod 26 = 12\bmod 26 = 12$	M
N	13	11	$(13+11)\bmod 26 = 24\bmod 26 = 24$	Y
G	6	8	$(6+8)\bmod 26 = 14\bmod 26 = 14$	O
W	22	13	$(22+13)\bmod 26 = 35\bmod 26 = 9$	J
E	4	6	$(4+6)\bmod 26 = 10\bmod 26 = 10$	K
L	11	15	$(11+15)\bmod 26 = 26\bmod 26 = 0$	A
L	11	4	$(11+4)\bmod 26 = 15\bmod 26 = 15$	P

The encrypted message would be XQ RZB SKTPMYO JKAP.

✐ Question 3

Given that CUTE was used as a key word for the Vigenère cipher to encrypt OUML KM CSA, what is the decrypted message?

Answer

Math is joy.

⚸⊸ Key idea

In a **Cryptogram** one letter stands for another. By examining the frequency that a letter occurs, one can try to decipher it.

⚸⊸ Key idea

Cable television companies verify your key as a valid customer before unscrambling the signal. They use a method of matching strings by "addition": $0 + 0 = 0$, $0 + 1 = 1$, $1 + 1 = 0$. Two strings match if they "add" to a string of 0's.

⌇ Example G

If your **key** is $k = 10011101$ and the transmitted message is $p + k = 01101100$, what is the password p that will unscramble your signal?

Solution

Add the strings $10011101 + 01101100$ to get 11110001.

$$
\begin{array}{r}
10011101 \\
+01101100 \\
\hline
11110001
\end{array}
$$

⚸⊸ Key idea

The **multiplication property for modular arithmetic** is as follows.

$$(ab)\bmod n = ((a\bmod n)(b\bmod n))\bmod n$$

This property allows for simpler calculations.

⌇ Example H

Use the multiplication property for modular arithmetic to simply $(11^2 \cdot 12)\bmod 5$.

Solution

$$
\begin{aligned}
(11^2 \cdot 12)\bmod 5 &= \left[(11^2 \bmod 5)(12\bmod 5)\right]\bmod 5 \\
&= \left[(11\bmod 5)^2 (2)\right]\bmod 5 \\
&= \left[(1^2 \bmod 5)(2)\right]\bmod 5 \\
&= \left[(1\bmod 5)(2)\right]\bmod 5 \\
&= \left[1(2)\right]\bmod 5 \\
&= 2\bmod 5 \\
&= 2
\end{aligned}
$$

⚿ Key idea

RSA public key encryption involves a procedure involving prime numbers and modular arithmetic. The procedure for sending and receiving messages is outlined on pages 642 – 643 of your text. In this form of encryption, letters start with 01 and a space is 00.

Letter	A	B	C	D	E	F	G	H	I	J	K	L	M
Location	01	02	03	04	05	08	07	08	09	10	11	12	13

Letter	N	O	P	Q	R	S	T	U	V	W	X	Y	Z
Location	14	15	16	17	18	19	20	21	22	23	24	25	26

⌒ Example I

Use the RSA scheme with $p = 7$, $q = 11$, and $r = 7$ to decode the received numbers 31, 01 and 48.

Solution

1. Since $p = 7$ and $q = 11$, $n = pq = 7 \cdot 11 = 77$.
2. Since $p - 1 = 7 - 1 = 6$ and $q - 1 = 11 - 1 = 10$, m will be the least common multiple of 6 and 10, namely 30.
3. We need to choose r such that it has no common divisors with 30. Thus, r can be 7. This confirms that $r = 7$ is a valid choice.
4. We need to find s.

$$r^2 \bmod m = 7^2 \bmod 30 = 49 \bmod 30 = 19$$

$$r^3 \bmod m = 7^3 \bmod 30 = 343 \bmod 30 = (11 \cdot 30 + 13) \bmod 30 = 13$$

$$r^4 \bmod m = 7^4 \bmod 30 = 2401 \bmod 30 = (80 \cdot 30 + 1) \bmod 30 = 1$$

Thus $t = 4$.

Since $s = r^{t-1} \bmod m$, where $r = 7$, $m = 30$, and $t = 4$, we have the following.

$$s = 7^{4-1} \bmod 30 = 7^3 \bmod 30 = 13$$

5. Since $31^{13} \bmod 77 = \left[\left(31^4 \right)^2 \cdot 31^5 \right] \bmod 77 = \left[\left(\left(31^4 \right) \bmod 77 \right)^2 \cdot 31^5 \bmod 77 \right] \bmod 77$

$$= \left(60^2 \cdot 12 \right) \bmod 77 = 43200 \bmod 77 = 3, \; R_1 = 3.$$

Since $1^{13} \bmod 77 = 1 \bmod 77 = 1$, $R_2 = 1$.

Since $48^{13} \bmod 77 = \left[\left(48^4 \right)^2 \cdot 48^5 \right] \bmod 77 = \left[\left(\left(48^4 \right) \bmod 77 \right)^2 \cdot 48^5 \bmod 77 \right] \bmod 77$

$$= \left(36^2 \cdot 34 \right) \bmod 77 = 44064 \bmod 77 = 20, \; R_3 = 20.$$

Thus, the message is CAT.

✐ Question 4

Use the RSA scheme with $p = 13$, $q = 11$, and $r = 7$ to encrypt the message DOG.

Answer

Thus, the numbers sent are 82, 115, 6.

Section 17.4   Web Searches and Mathematical Logic

☞ Key idea

Information on the Internet is stored on computers all around the world in the form of documents called Web pages. To find a particular source of information, web search engines are used to filter through web pages worldwide. These search engines use systematic search techniques called algorithms to gather information. When a search engine conducts a search, it must present the results in some sort of logical order. Many search engines rank each Web page according to features such as frequency with which the key word appears on the page or by the number of other pages that link to each page that is found.

☞ Key idea

An expression in **Boolean logic** is simply a statement that is either true or false. Many search engines allow users to construct searches using Boolean logic.

☞ Key idea

A **truth table** lists the values for an expression for all possible combinations of the Boolean variables P and Q. The letter T is used to indicate that an expression is true, and the letter F indicates that an expression is false.

☞ Key idea

The **connectives** AND, OR, and NOT can be used to construct complex Boolean expressions.

- The expression NOT P is called the **negation** of P. If P is true then NOT P is false, and if P is false then NOT P is true. The standard mathematical notation for this is $\neg P$. The truth table is as follows.

P	$\neg P$
T	F
F	T

- The expression P AND Q is called the **conjunction** of P and Q. This expression is true when both P and Q are true and is false otherwise. The standard mathematical notation for this is $P \wedge Q$. The truth table is as follows.

P	Q	$P \wedge Q$
T	T	T
T	F	F
F	T	F
F	F	F

- The expression P OR Q is called the **disjunction** of P and Q. This expression is true if either P or Q (or both) are true and is false otherwise. The standard mathematical notation for this is $P \vee Q$. The truth table is as follows.

P	Q	$P \vee Q$
T	T	T
T	F	T
F	T	T
F	F	F

🔑 Key idea

If two expressions have the same value (true or false) for each possible assignment of the Boolean variables, they are said to be **logically equivalent**.

Example J

Is the expression $P \vee (\neg Q \wedge R)$ logically equivalent to $(P \vee \neg Q) \wedge (P \vee R)$?

Solution

First we construct the truth table for $P \vee (\neg Q \wedge R)$.

P Q R	$\neg Q$	$\neg Q \wedge R$	$P \vee (\neg Q \wedge R)$
T T T	F	F	T
T T F	F	F	T
T F T	T	T	T
T F F	T	F	T
F T T	F	F	F
F T F	F	F	F
F F T	T	T	T
F F F	T	F	F

Next we construct the truth table for $(P \vee \neg Q) \wedge (P \vee R)$.

P Q R	$\neg Q$	$P \vee \neg Q$	$P \vee R$	$(P \vee \neg Q) \wedge (P \vee R)$
T T T	F	T	T	T
T T F	F	T	T	T
T F T	T	T	T	T
T F F	T	T	T	T
F T T	F	F	T	F
F T F	F	F	F	F
F F T	T	T	T	T
F F F	T	T	F	F

Since the last columns of the two truth tables are identical, the expression $P \vee (\neg Q \wedge R)$ is logically equivalent to $(P \vee \neg Q) \wedge (P \vee R)$. .

✏️ Question 5

Is the expression $P \wedge (\neg Q \vee \neg R)$ logically equivalent to $(P \wedge \neg Q) \vee (P \wedge \neg R)$?

Answer

Yes

Homework Help

Exercises 1 – 6

Carefully read Section 17.1 before responding to these exercises. The following diagrams may be helpful.

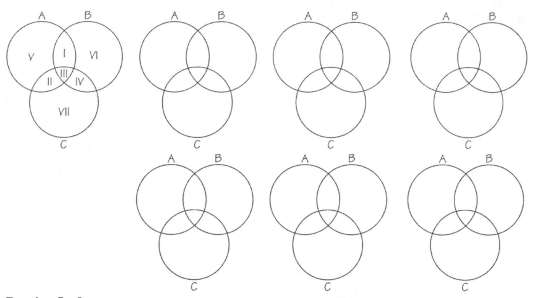

Exercises 7 – 8

Carefully read Section 17.2 before responding to these exercises. The following table may be helpful for Exercise 7.

	$a_2 + a_3$	c_1	$a_1 + a_3$	c_2	$a_1 + a_2$	c_3	Code word
000							
100							
010							
001							
110							
101							
011							
111							

Exercises 9 – 10

Carefully read Section 17.2 before responding to these exercises. The following table may be helpful for Exercise 9.

	$a_2 + a_3 + a_4$	c_1	$a_2 + a_4$	c_2	$a_1 + a_2 + a_3$	c_3	Code word
0000							
1000							
0100							
0010							
0001							
1100							
1010							
1001							
0110							
0101							
0011							
1110							
1101							
1011							
0111							
1111							

Exercise 11

Carefully read Section 17.2 before responding to this exercise. The following table may be helpful.

	$a_1 + a_2$	c_1	$a_2 + a_3$	c_2	$a_1 + a_3$	c_3	Code word
000							
100							
010							
001							
110							
101							
011							
111							

Exercise 12
Carefully read Section 17.2 before responding to this exercise. The following table may be helpful.

	Weight	Append	Code word
0000000			
0001011			
0010111			
0100101			
1000110			
1100011			
1010001			
1001101			
0110010			
0101110			
0011100			
1110100			
1101000			
1011010			
0111001			
1111111			

Exercise 13
Carefully read Section 17.2 before responding to this exercise. The following table may be helpful.

	Weight	Append	Code word
0000000			
0001011			
0010111			
0100101			
1000110			
1100011			
1010001			
1001101			
0110010			
0101110			
0011100			
1110100			
1101000			
1011010			
0111001			
1111111			

Exercises 14 – 16
Carefully read Section 17.2 before responding to these exercises. The following table may be helpful

Exercises 17 – 19
Carefully read Section 17.2 before responding to these exercises. The following tables may be helpful for Exercises 17 and 18.

	$a_1 + a_2$	$c_1 = (a_1 + a_2) \bmod 3$	$2a_1 + a_2$	$c_2 = (2a_1 + a_2) \bmod 3$	Code word
00					
10					
20					
01					
02					
11					
22					
21					
12					

Code word	1211 differs by

Exercises 21 – 24
Carefully read Section 17.2 before responding to these exercises. Pay special attention to Examples 1 and 2 on pages 630 – 631.

Exercises 25 – 28
Carefully read Section 17.2 before responding to these exercises. The following copy of the Morse code should be helpful for Exercises 25 – 27. In Exercise 28, answers will vary.

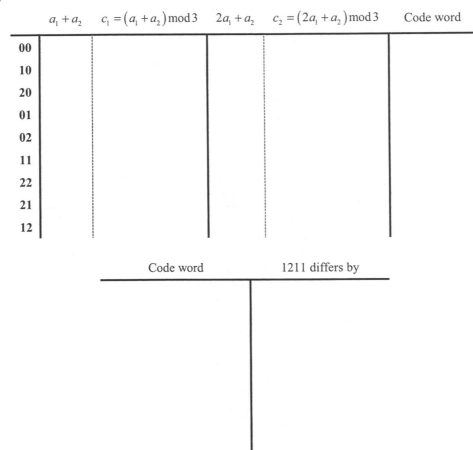

Exercises 29 – 30
Carefully read Section 17.2 before responding to these exercises. The following tables should be helpful.

Exercise 29

2015	2015
2057 – 2015	

Exercise 30

1207	1207
1207 + 373	

Exercises 31 – 33
Carefully read Section 17.2 before responding to these exercises. Pay special attention to Huffman coding on pages 631 – 634.

Exercises 34 – 37
Carefully read Section 17.3 before responding to these exercises. Pay special attention to modular arithmetic and the Caesar cipher on pages 634 – 636.

Exercises 38 – 40
Carefully read Section 17.3 before responding to these exercises. Pay special attention to Example 3 on pages 636 – 637. The following tables may be helpful.

For Exercise 38

Original	Location			Encrypted
			( ) mod 26 = mod 26 =	
			( ) mod 26 = mod 26 =	
			( ) mod 26 = mod 26 =	
			( ) mod 26 = mod 26 =	
			( ) mod 26 = mod 26 =	
			( ) mod 26 = mod 26 =	
			( ) mod 26 = mod 26 =	
			( ) mod 26 = mod 26 =	
			( ) mod 26 = mod 26 =	

For Exercise 39

Encrypted	Location				Decrypted
			= mod 26 = ( ) mod 26		
			= mod 26 = ( ) mod 26		
			= mod 26 = ( ) mod 26		
			= mod 26 = ( ) mod 26		
			= mod 26 = ( ) mod 26		
			= mod 26 = ( ) mod 26		
			= mod 26 = ( ) mod 26		
			= mod 26 = ( ) mod 26		
			= mod 26 = ( ) mod 26		
			= mod 26 = ( ) mod 26		
			= mod 26 = ( ) mod 26		
			= mod 26 = ( ) mod 26		
			= mod 26 = ( ) mod 26		

Continued on next page

For Exercise 40

Original	Location			Encrypted
			( )mod 26 = mod 26 =	
			( )mod 26 = mod 26 =	
			( )mod 26 = mod 26 =	
			( )mod 26 = mod 26 =	
			( )mod 26 = mod 26 =	
			( )mod 26 = mod 26 =	
			( )mod 26 = mod 26 =	
			( )mod 26 = mod 26 =	
			( )mod 26 = mod 26 =	
			( )mod 26 = mod 26 =	
			( )mod 26 = mod 26 =	
			( )mod 26 = mod 26 =	
			( )mod 26 = mod 26 =	
			( )mod 26 = mod 26 =	
			( )mod 26 = mod 26 =	
			( )mod 26 = mod 26 =	

Exercises 41 – 47
Carefully read Section 17.3 before responding to these exercises. Pay special attention to pages 637 – 639.

Exercise 48 – 50, 58
Visit the Web addresses indicated in these exercise.

Exercises 51 – 52
Carefully read Section 17.3 before responding to these exercises. Pay special attention to Example 4 on page 640.

Exercises 53 – 57

Carefully read Section 17.3 before responding to these exercises. Pay special attention to the procedure outlined on pages 642 – 643. A calculator is needed to do modular arithmetic with large values. For example $23^{13} \bmod 29 = \left[\left(23^4\right)^2 \cdot 23^5 \right] \bmod 29 = \left[\left(\left(23^4\right) \bmod 29 \right)^2 \cdot 23^5 \bmod 29 \right] \bmod 29$

$23^4 \bmod 29 = 20$ $23^5 \bmod 29$

$$23^{13} \bmod 29 = \left(20^2 \cdot 25\right) \bmod 29 = 10000 \bmod 29 = 24$$

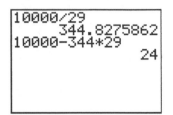

Exercises 59 – 67

Carefully read Section 17.4 before responding to these exercises. For exercises that require creating truth tables, the following should be helpful.

P	Q	R	
T	T	T	
T	T	F	
T	F	T	
T	F	F	
F	T	T	
F	T	F	
F	F	T	
F	F	F	

Do You Know the Terms?

Cut out the following 22 flashcards to test yourself on Review Vocabulary. You can also find these
flashcards at http://www.whfreeman.com/fapp7e.

Chapter 17 **Information Science** ## Binary linear code	**Chapter 17** **Information Science** ## Boolean Logic
Chapter 17 **Information Science** ## Caesar cipher	**Chapter 17** **Information Science** ## Cryptogram
Chapter 17 **Information Science** ## Cryptography	**Chapter 17** **Information Science** ## Data compression
Chapter 17 **Information Science** ## Decoding	**Chapter 17** **Information Science** ## Distance between two strings

Logic attributed to that uses operations such as $\wedge, \vee$, and $\neg$ to connect statements.	A code consisting of words composed of 0's and 1's obtained by using parity-check sums to append check digits to messages.
A sentence (or message) that has been encrypted.	A cryptosystem used by Julius Caesar whereby each letter is shifted the same amount.
The process of encoding data so that the most frequently occurring data are represented by the fewest symbols.	The study of how to make and break secret codes.
The distance between two strings of equal length is the number of positions in which they differ.	The process of translating received data into code words.

Chapter 17 Information Science **Encryption**	Chapter 17 Information Science **Even parity**
Chapter 17 Information Science **Key**	Chapter 17 Information Science **Key word**
Chapter 17 Information Science **Logically equivalent**	Chapter 17 Information Science **Modular arithmetic**
Chapter 17 Information Science **Nearest-neighbor decoding**	Chapter 17 Information Science **Odd parity**

Even integers are said to have even parity.	The process of encoding data to protect against unauthorized interpretation.
A word used to determine the amount of shifting for each letter while encoding a message.	A string used to encode and decode data.
Addition and multiplication involving modulo n.	Two expressions are said to be logically equivalent if they have the same values for all possible values of their Boolean variables.
Odd integers are said to have odd parity.	A method that decodes a received message as the code word that agrees with the message in the most positions.

Chapter 17
Information Science

Parity-check sums

Chapter 17
Information Science

RSA public key encryption scheme

Chapter 17
Information Science

Truth table

Chapter 17
Information Science

Variable-length code

Chapter 17
Information Science

Vignenère code

Chapter 17
Information Science

Weight of a binary code

A method of encoding that permits each person to announce publicly the means by which secret messages are to be sent to him or her.

Sums of digits whose parities determine the check digits.

A code in which the number of symbols for each code word may vary.

A tabular representation of an expression in which the variables and the intermediate expressions appear in columns and the last column contains the expression being evaluated.

The minimum number of 1's that occur among all nonzero code words of a code.

A cryptosystem that utilizes a key word to determine how much each letter is shifted.

Practice Quiz

1. If you use the circular diagram method to encode the message 1010, what is the encoded message?

   a. 1010001

   b. 1010010

   c. 1101000

2. Suppose the message 1111100 is received and decoded using the nearest-neighbor method. What message is recovered?

   a. 1111

   b. 1110

   c. 0111

3. What is the distance between received words 1111100 and 1101101?

   a. 5

   b. 4

   c. 2

4. For the code $C = \{0000, 1010, 0101, 1111\}$, how many errors would have to occur during the transmission for a received word to possibly be encoded incorrectly?

   a. 1

   b. 2

   c. 3

5. Use the encoding scheme $A \rightarrow 0$, $B \rightarrow 10$, $C \rightarrow 11$ to decode the sequence 0110011.

   a. ABBAC

   b. ACBAC

   c. ACAAC

6. What is the sum of the binary sequences 1001101 and 1100011?

   a. 1101111

   b. 0101110

   c. 0110000

7. Using modular arithmetic, $3^6 \bmod 20$

   a. 3.

   b. 9.

   c. 19.

8. For the RSA scheme with $p = 11$ and $q = 17$, which of the following could be chosen as a value for r?

   a. 5

   b. 6

   c. 9

9. For the RSA scheme with $m = 5$ and $r = 8$, what is the value of s?

   a. 2

   b. 3

   c. 5

10. Are $\neg(P \vee \neg Q)$ and $\neg P \vee \neg Q$ logically equivalent?

   a. no

   b. yes

   c. can't tell

Word Search

Refer to page 652 of your text to obtain the Review Vocabulary. There are 22 hidden vocabulary words/expressions in the word search below. This represents all vocabulary. It should be noted that spaces and hyphens are removed as well as accents.

```
T N A N Y G E R P S J E I A W E Z E O Y N U B R L H P U
F R U B I N A R Y L I N E A R C O D E H A F F P M O S Y
O I J I R F Z O S P T S S M O N O I T P Y R C N E L D A
O Y R P Z E N M P N E P O T G H A S P P E M E E B A P A
I B E D A Q N C C S Y B S J E P T T E S L K P T L E C F
T A G R T N E L A V I U Q E Y L L A C I G O L A H U D A
E N A N S U X Y Q E K O K F H V E N H E U I D D E S E E
U P J R E L E N O I S S E R P M O C A T A D E S A I C U
N S E G N I D O C E D R O B H G I E N T S E R A E N O F
S P D M M G E I T N J Q O M R L O B V D H C I O M A D E
E D O C Y R A N I B A F O T H G I E W O O A B N P X I T
R C C S A G N O E Y C U T R U T H T A B L E M M J Q N B
W E E A G I F A O D V E E N R L M W S C E D D M N N G F
G L R N H G I E D O C H T G N E L E L B A I R A V P A E
E M E H C S N O I T P Y R C N E Y E K C I L B U P A S R
C G N F I I I F D I S X V R E W Y N S L O I A M C E B E
D G E G D C I I C D S C I T E M H T I R A R A L U D O M
D D N G S Y S R G E T O W G E J Y W I I F R V X D M O P
R B G E W D A E P I T R A R L E Y O C I A N T Y Y A L E
H E I O P S R I B E U L E K K N F S F K M A G N H R E C
U R V S E P X S M U S K C E H C Y T I R A P D D O G A I
W N W A E V E N P A R I T Y H P A R G O T P Y R C O N O
E J C Z M G R L S E K F T W G D R I G Z I T H C I T L O
J K H B D J L S A E M D S O T R Y N X M S N S S T P O T
H H F S G V I G D C Q H E R H W I G A M T S T R I Y G F
N E G M S P A H V F W N S D Y E R S S T V V E Q K R I E
S C U G I H S E O E F X E P I S D B P E S C T G H C C O
R S C S E A R X O P G S G D A K A F P W J X D V A L A N
```

1. _____
2. _____
3. _____
4. _____
5. _____
6. _____
7. _____
8. _____
9. _____
10. _____
11. _____
12. _____
13. _____
14. _____
15. _____
16. _____
17. _____
18. _____
19. _____
20. _____
21. _____
22. _____

Chapter 18
Growth and Form

Chapter Objectives

Check off these skills when you feel that you have mastered them.

☐ Determine the scaling factor when given the original dimensions of an object and its scaled dimensions.

☐ Given the original dimensions of an object and its scaling factor, determine its scaled dimensions.

☐ Calculate the change in area of a scaled object when its original area and the scaling factor are given.

☐ Calculate the change in volume of a scaled object when its original volume and the scaling factor are given.

☐ Determine whether two given geometric objects are similar.

☐ Locate on a number line the new location of a scaled point.

☐ When given the two-dimensional coordinates of a geometric object, its center, and the scaling factor, calculate its new coordinates after the scaling has taken place.

☐ Calculate from given formulas the perimeter and area of a two-dimensional object.

☐ Calculate from given formulas the surface area and volume of a three-dimensional object.

☐ Describe the concept of area-volume tension.

☐ Explain why objects in nature are restricted by a potential maximum size.

Guided Reading

Introduction

We examine how a variety of physical dimensions of an object – length, area, weight, and so on – are changed by proportional growth of the object. These changes influence the growth and development of an individual organism and the evolution of a species.

Section 18.1   Geometric Similarity

⌐ Key idea

Two objects are **geometrically similar** if they have the same shape, regardless of their relative sizes. The **linear scaling factor** relating two geometrically similar objects A and B is the ratio of the length of a part of B to the length of the corresponding part of A.

☞ Example A

Compare a cube A, which is 2 inches on a side, to a similar cube B, which is 12 inches on a side. What is the scaling factor of the enlargement from A to B and what is the ratio of the diagonal of B to the diagonal of A?

Solution

The scaling factor r is the ratio of any two corresponding linear dimensions. Thus r = ratio of sides = $\frac{12}{2} = 6$, which is also the ratio of diagonals when comparing the second object to the original.

⌐ Key idea

The area of the surface of a scaled object changes according to the square of the linear scaling factor.

☞ Example B

Compare a cube A, which is 2 inches on a side, to a similar cube B, which is 12 inches on a side. What is the ratio of the surface area of the large cube B to that of the small cube A?

Solution

You do not need to calculate the surface areas of the cubes. Area scales according to the square of the scaling factor. Since $r = 6$, $r^2 = 36$, and so we get $\dfrac{\text{area } B}{\text{area } A} = 6^2 = 36$.

⌐ Key idea

The volume (and weight) of a scaled object changes according to the cube of the linear scaling factor.

☞ Example C

Cube A, which is 2 inches on a side, is similar to cube B, which is 12 inches on a side. If the small cube A weighs 3 ounces, how much does the large cube B weigh?

Solution

Weight, like volume, scales by the cube of the scaling factor. Since $\dfrac{\text{weight } B}{\text{weight } A} = 6^3 = 216$, weight B

$= 216 \times 3$ oz $= 648$ oz, or 40.5 pounds.

✏ Question 1

A model of a water tower is built to a scale of 1 to 59. If the model holds 8 cubic inches, how much will the actual water tower hold?

Answer

Approximately 951 cubic feet

⚷ Key idea

In describing a growth situation or a comparison between two similar objects or numbers, the phrase "x is increased by" a certain percentage means that you must add the amount of the increase to the current value of x. The phrase "x is decreased by" a percentage means that you must subtract the amount of decrease from x.

↷ Example D

a)   This year, the value of the stock of the ABC Corporation rose by 25%, from 120 to ____.

b)   This year, the value of the stock of the XYZ Corporation fell by 25%, from 150 to ____.

Solution

a)   The new stock value is $\left(1+\frac{25}{100}\right)\times120=120+30=150$.

b)   The new stock value is $\left(1-\frac{25}{100}\right)\times150=150-37.5=112.5$.

✏ Question 2

a)   This year, the value of the stock of the Dippy Dan Corporation fell by x%, from 100 to 80. What is x?

b)   This year, the value of the stock of the Buckaroo Corporation rose by x%, from 80 to 100. What is x?

Answer

a)   20
b)   25

Section 18.2   How Much Is That in ...?

⚷ Key idea

Some basic dimensional units in the US system of measurement are: foot (length), gallon (volume), pound (weight). Some comparable units in the metric system are: meter (length), liter (volume), kilogram (weight). Refer to Table 18.3 on page 671 for conversion between the two measuring systems.

⚷ Key idea

Conversions from one system to the other are done according to our rules for scaling. For example, 1 meter = 3.28 feet; that is, we have a scaling factor of 3.28 from meters to feet. Therefore, an area of $5\text{ m}^2 = 5\times(3.281)^2\text{ ft}^2 \approx 53.82\text{ ft}^2$.

↷ Example E

Convert
a)   5 pounds = _____ kilograms
b)   60 kilometers = _____ miles
c)   2000 in^2 = _____ m^2

Solution

These answers are approximate.
a)   1 lb ≈ 0.4536 kg, so 5 lb = 5 × 0.4536 = 2.27 kg.
b)   1 km = 0.621 mi, so 60 km = 60 × 0.621 = 37.26 mi.
c)   1 in = 2.54 cm = 0.0254 m, so 1 in^2 = $(0.0254)^2$ = 0.000645 m^2;
     therefore, 2000 in^2 = 2000 × 0.000645 = 1.29 m^2.

✐ Question 3

Convert
a)   17 miles = _____ kilometers
b)   100 meters = _____ yards
c)   300 m^2 = _____ in^2

Answer

These answers are approximate.
a)   27.37
b)   109.4
c)   464,999

Section 18.3   Scaling a Mountain

☞ Key idea

The size of a real object or organism is limited by a variety of structural or physiological considerations. For example, as an object is scaled upward in size, the **mass** and weight grow as the cube of the scaling factor, whereas, the surface area grows as the square of the same factor.

☞ Key idea

The **pressure** on the bottom face or base of an object (for example, the feet of an animal or the foundation of a building) is the ratio of the weight of the object to the area of the base; that is,

$P = \dfrac{W}{A}$. The weight may be calculated by multiplying the volume by the density.

↷ Example F

What is the pressure at the base of a block of stone that is 2 ft wide, 3 ft long and 4 ft high, given that the density of the stone is 350 lb per ft^3?

Solution

$P = \dfrac{W}{A}$. The weight W of the block is volume × density.

Thus, W = (2 ft × 3 ft × 4 ft) × 350 lb/ft^3 = 8400 lb.
The base area A = 2 × 3 = 6 ft^2. Therefore, the pressure P = 8400/6 = 1400 lb/ft^2.

⚕ Example G

What are the dimensions and the pressure at the base of a block of the same stone that is the same shape but scaled up by a factor of 5?

Solution

The scaling factor is 5. Volume, and therefore weight, scales by $5^3 = 125$, while area scales by $5^2 = 25$. Thus, the pressure is $P = \dfrac{125 \times W}{25 \times A} = \dfrac{125 \times (24 \times 350)}{25 \times 6} = 7000\,\text{lb/ft}^2$.

⚕ Example H

What is the pressure at the base of a cylindrical column of the same stone that is 3 feet in diameter and 60 feet high, if the density of the stone is 350 lb/sq ft?

Solution

The diameter of the column is 3, so the radius is 1.5. The column is a cylinder, so its volume is $\pi \times (\text{radius})^2 \times \text{height} = 3.14 \times 2.25 \times 60 = 423.9\,\text{ft}^3$. Because the density is 350 lb/ft^3, the weight $W = \text{volume} \times \text{density} = 423.9 \times 350 = 148{,}365\,\text{lb}$. The area A of the base is $\pi \times (\text{radius})^2 = 7.065$ ft^2. Then $P = \dfrac{W}{A} = \dfrac{148{,}365}{7.065} = 21{,}000\,\text{lb/ft}^2$.

✎ Question 4

The weight of a block of marble that measures 1 ft×2 ft×3 ft weighs 1240 lbs. If 10 of these blocks are used to make a wall 6 ft high, 10 ft long, and 1 ft wide, what is the pressure on the bottom faces?

Answer

Approximately 8.61 lb/in^2

Section 18.4   Sorry, No King Kongs

⚖ Key idea

Though the weight of an object increases with the cube of the linear scaling factor, the ability to support weight increases only with the square of the linear scaling factor.

⚖ Key idea

As an object is scaled up in size, the area of its surface increases as the square of the scaling factor, while the volume increases as the cube of the scaling factor. This **area-volume tension** can strongly influence the development of structural parts of organisms that depend on both dimensions for strength, mobility, heat control, breathing, flight, and so on.

⚖ Key idea

Proportional growth does not preserve many organic properties. As individuals grow, or as species evolve, their physiological proportions must change; growth is not proportional.

Section 18.5   Dimension Tension

⌐ Key idea

Area-volume tension is a result of the fact that as an object is scaled up, the volume increases faster than the surface area and faster than areas of cross sections.

⌐ Key idea

Proportional growth is growth according to geometric similarity: the length of every part of the organism enlarges by the same linear scaling factor.

Proportional growth, scaling factor = 2

Disproportionate growth: height scaling factor = 2, length scaling factor = 4

Section 18.6   How to Grow

⌐ Key idea

Allometric growth is the growth of the length of one feature at a rate proportional to a power of the length of another.

⌐ Key idea

On a graph we can use **orders of magnitude**, such as powers of 10 $\left(10^0, 10^1, 10^2, ...\right)$, instead of integer values such as 0, 1, 2, ... When the scale is powers of 10, it is called **base-10 logarithmic scale**. On graph paper when both axes are scaled as such, the graph paper is called **log-log paper**. When only one axis is scaled as such, it is called **semilog paper**.

⌐ Key idea

Large changes in scale force a change in either material or form. Thus, limits are imposed on the scale of living organisms.

⌐ Key idea

A **power curve** is described by the equation $y = bx^a$. In this case y is proportional to x.

✍ Example I

The following is data concerning the growth of an ebix (fictitious animal).

Age	Height	Log (Height)	Nose length	Log (Nose length)
3.25	61 in	1.79	20 in	1.30
4.2	100 in	2.00	50 in	1.70

Find the slope of the line from ages 3.25 to 4.2, where the *height* is plotted on the horizontal axis and the *nose length* is on the vertical axis (log of each, on log-log paper).

Solution

The slope for the line from ages 3.25 to 4.2 is the vertical change over the horizontal change in terms of log units.

$$\frac{\log 50 - \log 20}{\log 100 - \log 61} = \frac{1.70 - 1.30}{2.00 - 1.79} = \frac{0.4}{0.21} \approx 1.9$$

Thus, the slope of the line on log-log paper is approximately 1.9.

Homework Help

Exercises 1 – 14
Carefully read Section 18.1 before responding to these exercises.

Exercises 15 – 23
Carefully read Section 18.2 before responding to these exercises.

Exercises 24 – 31
Carefully read Section 18.3 before responding to these exercises.

Exercises 32 – 45
Carefully read Section 18.4 before responding to these exercises.

Exercises 46 – 57
Carefully read Section 18.5 before responding to these exercises.

Exercises 58 – 59
Carefully read Section 18.6 before responding to these exercises.

Do You Know the Terms?

Cut out the following 18 flashcards to test yourself on Review Vocabulary. You can also find these flashcards at http://www.whfreeman.com/fapp7e.

Chapter 18 Growth and Form **Allometric growth**	Chapter 18 Growth and Form **Area-volume tension**
Chapter 18 Growth and Form **Base-10 logarithmic scale**	Chapter 18 Growth and Form **Crushing strength**
Chapter 18 Growth and Form **Density**	Chapter 18 Growth and Form **Dilation**
Chapter 18 Growth and Form **Geometrically similar**	Chapter 18 Growth and Form **Isometric growth**

A result of the fact that as an object is scaled up, the volume increases faster than the surface area and faster than areas of cross sections	A pattern of growth in which the length of one feature grows at a rate proportional to a power of the length of another feature
The maximum ability of a substance to withstand pressure without crushing or deforming	A scale on which equal divisions correspond to powers of 10
A linear scaling	Weight per unit volume
Proportional growth	Two objects are this if they have the same shape, regardless of the materials of which they are made. They need not be of the same size. Corresponding linear dimensions must have the same factor of proportionality

Chapter 18
Growth and Form

Linear scaling factor

Chapter 18
Growth and Form

Log-log paper

Chapter 18
Growth and Form

Orders of magnitude

Chapter 18
Growth and Form

Power curve

Chapter 18
Growth and Form

Pressure

Chapter 18
Growth and Form

Problem of scale

Chapter 18
Growth and Form

Proportional growth

Chapter 18
Growth and Form

Semilog paper

Graph paper on which both the vertical and the horizontal scales are logarithmic scales, that is, the scales are marked in orders of magnitude 1, 10, 100, 1000, ..., instead of 1, 2, 3, 4, . .	The number by which each linear dimension of an object is multiplied when it is scaled up or down; that is, the ratio of the length of any part of one of two geometrically similar objects to the length of the corresponding part of the second.
A curve described by an equation $y = bx^a$, so that y is proportional to a power of x	Powers of 10
As an object or being is scaled up, its surface and cross-sectional areas increase at a rate different from its volume, forcing adaptations of materials or shape.	Force per unit area
Graph paper on which only one of the scales is a logarithmic scale	Growth according to geometric similarity, where the length of every part of the organism enlarges by the same linear scaling factor

Chapter 18
Growth and Form

Weight

Chapter 18
Growth and Form

Wing loading

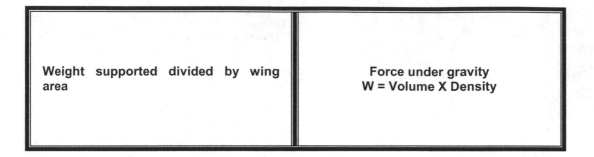

| Weight supported divided by wing area | Force under gravity
W = Volume X Density |

Practice Quiz

1. You want to enlarge a small painting, measuring 4 inches by 7 inches, onto a piece of paper, measuring 8.5 inches by 11 inches, so that the image is proportional and as large as possible. What is the scaling factor for the enlargement?

    a. 1.571

    b. 2.125

    c. 3.339

2. A small quilted wall hanging has an area of 15 sq. ft. A larger bed quilt is proportional to the wall hanging with a linear scaling factor of 3. What is the area of the bed quilt?

    a. 3 sq. ft.

    b. 45 sq. ft.

    c. 135 sq. ft.

3. A model airplane is built to a scale of 1 to 50. If the wingspan of the actual airplane is 25 ft, what is the wingspan of the model plane?

    a. 2 ft

    b. 1/2 ft

    c. 1/10 ft

4. At the grocery store, an 8-inch cherry pie costs $4.39 and a similar 10-inch cherry pie costs $6.15. Which is the better buy?

    a. the 8-inch pie

    b. the 10-inch pie

    c. they are about the same

5. A 3rd grade class glues 64 sugar cubes together to form a larger cube. What is the linear scaling factor?

    a. 4

    b. 8

    c. 64

6. 5 cm is approximately _____ in.

    a. 1.6

    b. 1.97

    c. 12.7

7. 9.5 sq. mi is approximately _____ sq. km.

    a. 15.30

    b. 3.66

    c. 24.62

8. A table weighs 125 pounds and is supported by four legs, which are each 0.75 inch by 0.5 inch by 24 inches high. How much pressure do the legs exert on the floor?

   a. 333 lb/sq. in

   b. 83 lb/sq in

   c. 2000 lb/sq in

9. The growth of human bodies is best modeled as _____.

   a. isometric growth

   b. proportional growth

   c. allometric growth

10. If points $(1,5)$ and $(2,15)$ lie on the graph of $\log y = B + a \log x$, what is the value of a?

   a. $\dfrac{\log 15 - \log 5}{\log 2 - \log 1}$

   b. $\dfrac{\log 15 - \log 2}{\log 5 - \log 1}$

   c. $\dfrac{\log 15 - \log 1}{\log 5 - \log 2}$

Word Search

Refer to page 694 of your text to obtain the Review Vocabulary. There are 17 hidden vocabulary words/expressions in the word search below. It should be noted that spaces are removed as well as hyphens. Also, *Base-10 logarithmic scale* does not appear in the word search.

```
D E E G I S R D N E I F L F H E I L E I P X H F J
W N S A P I K H T W O R G C I R T E M O S I C R S
O D V C I E E R T T G D S R A H L R I S L O P I I
V T Z E Y U M R C R G S W U U A T T M I E E H I D
M A E M E Z G K D G G T P S M O E X O I O M M N I
N O E O K H K N J L E O E H E N N S H N R I E O L
N M I O N T P R O P O R T I O N A L G R O W T H A
S S F P E W F G L A M A T N F D D Z D M S K F T T
U E A V M O N K W R E I E G C D G D Q I C E Y T I
S P L J D R O E E J T C J S D E Q S V L O E Y Q O
M U L I J G S Q I V R R N T P R E C L U L N E C N
D C D P A C Q E G X I D H R C S R A E E H I R G D
N P L R S I Q N H F C F S E M I L O G P A P E R M
B W A E S R R O T C A F G N I L A C S R A E N I L
M N T S S T M T Y E L L O G L O G P A P E R Q G J
L E O S M E J O O N L I W T O A L G L N C I I O A
D F S U T M R E A N Y X M H Z Z H W O E J Z H E V
E E F R Z O R D E R S O F M A G N I T U D E Q W E
D I D E C L E U N O I S N E T E M U L O V A E R A
R N A X H L U E E I M V A E G M H N S A F P M P M
E G N I D A O L G N I W F N R Y T I S N E D D A U
T E L A C S F O M E L B O R P O W E R C U R V E O
L I D I R O E T R I A S P I Y E M D T N B E E T N
Q S W R N A T C R Y R S A J C A K T B H J Z O P G
I B K O O P Y H G B A T E A M D Y H E U Z H E A S
```

1. _____

2. _____

3. _____

4. _____

5. _____

6. _____

7. _____

8. _____

9. _____

10. _____

11. _____

12. _____

13. _____

14. _____

15. _____

16. _____

17. _____

Chapter 19
Symmetry and Patterns

Chapter Objectives

Check off these skills when you feel that you have mastered them.

☐ List the first ten terms of the Fibonacci sequence.

☐ Beginning with the number 3, form a ratio of each term in the Fibonacci sequence with its next consecutive term and simplify the ratio; then identify the number that these ratios approximate.

☐ List the numerical ratio for the golden section.

☐ Name and define the four transformations (rigid motions) in the plane.

☐ Analyze a given rosette pattern and determine whether it is dihedral or cyclic.

☐ Given a rosette pattern, determine which rotations preserve it.

☐ Analyze a given strip pattern by determining which transformations produced it.

Guided Reading

Introduction

We study certain numerical and geometric patterns of growth and structure that can be used to model or describe an amazing variety of phenomena in mathematics and science, art, and nature. The mathematical ideas the Fibonacci sequence leads to, such as the golden ratio, spirals, and selfsimilar curves, have long been appreciated for their charm and beauty; but no one can really explain why they are echoed so clearly in the world of art and nature. The properties of selfsimilarity, and reflective and rotational symmetry are ubiquitous in the natural world and are at the core of our ideas of science and art.

⌗ **Key idea**

Plants exhibiting **phyllotaxis** have a number of spiral forms coming from a special sequence of numbers. Certain plants have spirals that are geometrically similar to one another. The spirals are arranged in a regular way, with balance and "proportion". These plants have **rotational symmetry**.

Section 19.1   Fibonacci Numbers and the Golden Ratio

⌗ **Key idea**

Fibonacci numbers occur in the sequence {1, 1, 2, 3, 5, 8, 13, 21, 34, 55, 89, . . . }. They are generated according to the **recursion** formula that states that each term is the sum of the two terms preceding it. If the nth Fibonacci number is F_n then for $F_1 = F_2 = 1$ and $n \geq 2$, we have the following.

$$F_{n+1} = F_n + F_{n-1}$$

🌊 Example A

Fill in the next five terms in the Fibonacci sequence $\{1, 1, 2, 3, 5, 8, 13, 21, 34, 55, 89, \ldots\}$.

Solution

Using the recursion formula we have the following.

$F_{12} = F_{11} + F_{10} = 89 + 55 = 144, \quad F_{13} = F_{12} + F_{11} = 144 + 89 = 233, \quad F_{14} = F_{13} + F_{12} = 233 + 144 = 377,$

$F_{15} = F_{14} + F_{13} = 377 + 233 = 610, \quad F_{16} = F_{15} + F_{14} = 610 + 377 = 987$

✏️ Question 1

Suppose a sequence begins with numbers 2 and 4, and continues by adding the previous two numbers to get the next number in sequence. What would the sum of the 10th and 11th terms in this sequence be?

Answer

466, the 12th term

🔑 Key idea

As you go further out in the sequence, the ratio of two consecutive Fibonacci numbers approaches the famous golden ratio $\phi = 1.618034\ldots$. For example, $89/55 = 1.61818\ldots$ and $377/233 = 1.618025\ldots$.

The number ϕ is also known as the golden mean. The exact value is $\phi = \dfrac{1+\sqrt{5}}{2}$.

🌊 Example B

Calculate the square of ϕ, subtract 1, and compare the result to ϕ. What do you get?

Solution

Since ϕ satisfies the algebraic equation $\phi^2 = \phi + 1$, $\phi^2 - 1 = \phi$.

🌊 Example C

Calculate the reciprocal $\dfrac{1}{\phi}$, add 1, and compare to ϕ. What do you get?

Solution

If you divide both sides of the equation $\phi^2 = \phi + 1$ by ϕ, you get $\phi = 1 + \dfrac{1}{\phi}$.

🔑 Key idea

A **golden rectangle** (which is considered by many to be visually pleasing) is one such that the ratio of height to width is 1 to ϕ.

$\phi = \dfrac{1+\sqrt{5}}{2}$

⌐ Key idea

The **geometric mean** of two positive numbers, a and b, is the square root of their product.

$$\sqrt{ab}$$

In general, the geometric mean of n numbers is the nth root of their product.

✎ Question 2

a)   What is the geometric mean of 5 and 7? (round to three decimal places)

b)   What is the geometric mean of 5, 6, and 7? (round to three decimal place)

Answer

a)   5.916

b)   5.944

Section 19.2   Symmetries Preserve the Pattern

⌐ Key idea

Balance refers to regularity in how repetitions are arranged. Along with **similarity** and **repetition,** balance is a key aspect of symmetry.

⌐ Key idea

Rigid motions are **translations, rotations, reflections,** and **glide reflections.**

⌐ Example D

Classify each of these rigid motions (within the given rectangle).

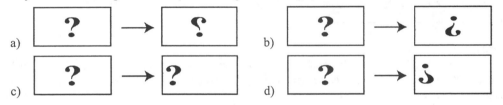

Solution

a)   This symmetry reverses left and right. This is a reflection across a vertical line.

b)   This reverses left-right and up-down. This is a rotation by 180°.

c)   The figure is moved to the left with the same orientation. This is a translation to the left.

d)   The figure is moved to the left and simultaneously reversed up-down. This is a glide reflection.

⌐ Key idea

Preservation of the pattern occurs when the pattern looks exactly the same, with all parts appearing in the same places, after a particular motion is applied.

✎ Question 3

Which of the following letters has a shape that is preserved by reflection or rotation?

A E I S

Answer

All four letteres

Section 19.3   Rosette, Strip, and Wallpaper Patterns

🔑 Key idea
Patterns are analyzed by determining which rigid motions preserve the pattern. These rigid motions are called **symmetries of the pattern**.

🔑 Key idea
Rosette patterns contain only rotations and reflections.

🔑 Key idea
Wallpaper patterns repeat a design element in more than one direction.

🔑 Key idea
Strip patterns repeat a design element along a line, so all of them have translation symmetry along the direction of the strip and may also contain glide reflections.

〰️ Example E
Which of the following linear designs are periodic strip patterns?

a)

b)

c)

Solution
In choice b, the design element consisting of two sets of bars is repeated at regular intervals. In choice a, there is no fixed repeated design element. In choice c, the intervals between repetitions are irregular. The answer is pattern b.

🔑 Key idea
Other possible symmetries for a strip pattern are horizontal or vertical reflection, rotation by 180°, or glide reflection.

Section 19.4   Notation for Patterns

🔑 Key idea
We can classify strip patterns according to their symmetry types; there are exactly seven different classes, designated by four symbols $p * * *$. The first symbol is always the p. The second symbol is either m or 1 indicating the presence or absence of a vertical line of reflection. The third symbol is m if there is a horizontal line of reflection, a if there is a glide reflection but no horizontal reflection, or 1 if there is horizontal or glide reflection. The fourth symbol is a 2 if there is half-turn rotational symmetry; otherwise it is a 1.

🖉 Example F

What is the symbolic notation for each of these patterns? You may need to refer to the flow chart (Figure 19.12) on page 729 of your text to help answer this question.

a)

b)

Solution

a) No vertical or horizontal reflection, or rotation. However, each figure matches the next one over with a horizontal flip; that is, it has glide reflection symmetry. This would be $p1a1$.

b) This design has all possible symmetries for strip patterns. This would be $pmm2$

⌐ Key idea

It is useful to have a standard notation for patterns, for purposes of communication. Crystallographer's notation is the one most commonly used.

⌐ Key idea

In applying notation to patterns, it must be taken into account that patterns may not be perfectly rendered, especially if they are on a rounded surface.

Section 19.5   Symmetry Groups

⌐ Key idea

A group is made up of a set of elements and an operation that has the following properties.

- Closure: The result of an operation on any two elements of the set yields an element of the set.

- Identity: The set has a special element I such that if the operation is performed with any element of the set, say A, the result of $A \circ I = I \circ A$, which is A.

- Inverse: For any element A, there exists an element of the set A^{-1} such that $A \circ A^{-1} = A^{-1} \circ A = I$.

- Associatively: For any three elements of the set, A, B, and C, we have the following.

$$A \circ B \circ C = A \circ (B \circ C) = (A \circ B) \circ C$$

⌐ Key idea

The full list of symmetries of any pattern forms a **symmetry group**. Symmetry of a pattern has the following properties.

- The combination of two symmetries A and B is written $A \circ B$. This combination is another symmetry.

- The "null" symmetry doesn't move anything. It is considered the identity.

- Every symmetry has an inverse or an opposite that "undoes" the effect of the original symmetry.

- In applying a number of symmetries one after another, we may combine consecutive ones without affecting the result.

⌐ Key idea

The symmetries of a rectangle are as follows.
- *I*: leaves every point unchanged in location
- *R*: A 180° half-turn through the center)
- *V*: A reflection in a vertical line through the center
- *H*: A reflection in a horizontal line through the center

⌐ Example G

What is $H \circ V$?

Solution

The answer is *R*. *H* reverses up and down, *V* reverses left and right. *R* reverses both, as does the combination symmetry $H \circ V$.

⌐ Example H

a)   List the four elements of the symmetry group of a rectangle.
b)   What is the combined result of applying all four of them consecutively?

Solution

a) The four elements that form the group are *I*, *H*, *V*, and *R*. A rectangle is symmetrical across a vertical and horizontal axis, but not across the diagonals. It is, therefore, also symmetrical if rotated 180°, since $H \circ V = R$.
b) To calculate the combination $I \circ V \circ H \circ R$, first note $I \circ V = V$. Then $V \circ H = R$, and finally $R \circ R = I$. Thus, the answer is *I*.

⌐ Key idea

Symmetry groups of rosette patterns contain only rotations and reflections.

⌐ Example I

What are the symmetries of these rosettes?

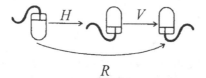

Solution

The answer is:

a)   Three rotations $\{I, R, R^2\}$, where *R* is a rotation 120°, and three reflections across the axes *a*, *b*, *v* shown below.

b)   Three rotations $\{I, R, R^2\}$, where *R* is a rotation 120°, but no reflections, since the figure has a "right-handed" twist, and any reflection would change it to a "left-handed" one.
c)   No symmetries other than *I*. The figure is totally asymmetric.

Section 19.6   Fractals Patterns and Chaos

⌗→ Key idea

A **fractal** is a kind of pattern that exhibits similarity at ever finer scales. When you "zoom in", the resulting figure looks like the original figure.

⌗→ Key idea

Fractals can be created with a replication rule called an **iterative function system** (**IFS**). One particular geometric pattern is called **Sierpinski's triangle**. It starts with a triangle and the "middle triangle" is removed. This process is repeated for the resulting triangles.

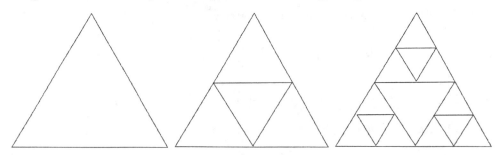

After a few more iterations the figure looks like the following.

This process continues.

✎ Question 4

A fractal known as the Koch snowflake starts with an equilateral triangle. Each side of the triangle is cut into three parts with the middle section removed. The removed segment is replaced by two line segments of the same length as that was removed. If the following is a picture of what occurs after doing this process twice, draw the first two pictures and the one that follows (the first is the equilateral triangle).

Is the area in the interior of the figure increasing as you do each iteration?

Answer

yes

Homework Help

Exercises 1 – 18
Carefully read Section 19.1 before responding to these exercises.

Exercises 19 – 26
Carefully read Section 19.2 before responding to these exercises.

Exercises 27 – 34
Carefully read Section 19.3 before responding to these exercises.

Exercises 35 – 48
Carefully read Section 19.4 before responding to these exercises.

Exercises 49 – 61
Carefully read Section 19.5 before responding to these exercises.

Exercises 62 – 68
Carefully read Section 19.6 before responding to these exercises.

Do You Know the Terms?

Cut out the following 23 flashcards to test yourself on Review Vocabulary. You can also find these flashcards at http://www.whfreeman.com/fapp7e.

Chapter 19 Symmetry and Patterns **Divine proportion**	Chapter 19 Symmetry and Patterns **Fibonacci numbers**
Chapter 19 Symmetry and Patterns **Fractal**	Chapter 19 Symmetry and Patterns **Geometric mean**
Chapter 19 Symmetry and Patterns **Glide reflection**	Chapter 19 Symmetry and Patterns **Golden ratio, golden mean**
Chapter 19 Symmetry and Patterns **Golden rectangle**	Chapter 19 Symmetry and Patterns **Generated, generators**

The numbers in the sequence 1, 1, 2, 3, 5, 8, 13, 21, 34, . . . (each number after the second is obtained by adding the two preceding numbers).	Another term for the golden ratio.
The geometric mean of two numbers a and b is $\sqrt{ab}$.	A pattern that exhibits similarity at ever-finer scales.
The number $\phi = \dfrac{1+\sqrt{5}}{2} = 1.618034. \ldots$	A combination of translation (= glide) and reflection in a line parallel to the translation direction.
A group is generated by a particular set of elements if composing them and their inverse in combinations can produce all elements of the group.	A rectangle the lengths of whose sides are in the golden ratio.

Chapter 19
Symmetry and Patterns

Group

Chapter 19
Symmetry and Patterns

Isometry

Chapter 19
Symmetry and Patterns

Iterated function system (IFS)

Chapter 19
Symmetry and Patterns

Phyllotaxis

Chapter 19
Symmetry and Patterns

Preserves the pattern

Chapter 19
Symmetry and Patterns

Recursion

Chapter 19
Symmetry and Patterns

Rigid motion

Chapter 19
Symmetry and Patterns

Rosette pattern

Another word for rigid motion. Angles and distances, and consequently shape and size, remain unchanged by a rigid motion.	A collection of elements with an operation on pairs of them such that the collection is closed under the operation, there is an identity for the operation, each element has an inverse, and the operation is associative.
The spiral pattern of shoots, leaves, or seeds around the stem of a plant.	A sequence of elements (number or figures) in which each successive element is determined recursively by applying the same function (rule) to the previous element.
A method of defining a sequence of numbers, in which the next number is given in terms of previous ones.	A transformation does this if all parts of the pattern look exactly the same after the transformation has been performed.
A pattern whose only symmetries are rotations about a single point and reflections through that point.	A motion that preserves the size and shape of figures; in particular, any pair of points is the same distance apart after the motion as before.

Chapter 19
Symmetry and Patterns

Rotational symmetry

Chapter 19
Symmetry and Patterns

Strip pattern

Chapter 19
Symmetry and Patterns

Symmetry of the pattern

Chapter 19
Symmetry and Patterns

Symmetry group of the pattern

Chapter 19
Symmetry and Patterns

Translation

Chapter 19
Symmetry and Patterns

Translation symmetry

Chapter 19
Symmetry and Patterns

Wallpaper pattern

A pattern that has indefinitely many repetitions in one direction.

A figure has this symmetry if a rotation about its "center" leaves it looking the same.

The group of symmetries that preserve the pattern.

A transformation of a pattern is this if it preserves the pattern.

An infinite figure has this symmetry if it can be translated (slid, without turning) along itself without appearing to have changed.

A rigid motion that moves everything a certain distance in one direction.

A pattern in the plane that has indefinitely many repetitions in more than one direction.

Practice Quiz

1. The numbers 21 and 34 are consecutive numbers in the Fibonacci sequence. What is the next Fibonacci number after 34?

   a. 46

   b. 55

   c. 59

2. What is the geometric mean of 8 and 32?

   a. 20

   b. 16

   c. 6.32

3. The shorter side of a golden rectangle is 7 inches. How long is the longer side?

   a. 8.6 inches

   b. 9.5 inches

   c. 11.3 inches

4. Which figure has rotation symmetry?

   I.   II. ⊔+⊔

   a. I only

   b. II only

   c. Both I and II

5. Assume the following patterns continue in both directions. Which has a reflection isometry?

   I. TTTTTTTTTTTTT   II. >>>>>>>

   a. I only

   b. II only

   c. Both I and II

6. Assume the following two patterns continue in both directions. Which of these patterns has a glide reflection isometry?

   I. ◇◇◇◇◇◇◇   II. ZZZZZZZZZZZ

   a. I only

   b. II only

   c. Neither I nor II

7. Use the flowchart in Figure 19.11 of *For All Practical Purposes* to identify the notation for the strip pattern below.

   ⇑⇓⇑⇓⇑⇓⇑⇓⇑⇓

   a. pma2

   b. p1a1

   c. p112

8.   What isometries does this wallpaper pattern have?

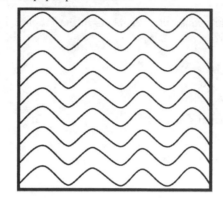

    a. translation only

    b. translation and reflection only

    c. translation, reflection, and rotation

9.   What isometries does this wallpaper pattern have?

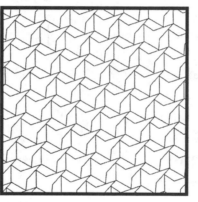

    a. translation only

    b. translation and reflection only

    c. translation, reflection, and rotation

10. How many elements are in the symmetry group of a regular hexagon?

    a. 6

    b. 12

    c. more than 12

Word Search

Refer to pages 740 – 741 of your text to obtain the Review Vocabulary. There are 25 hidden vocabulary words/expressions in the word search below. *Golden ratio*, *Golden mean*, *Generated*, *Generators*, *Translation*, and *Translation symmetry* all appear separately. It should be noted that spaces are removed. Also, abbreviations do not appear. There is additional space on the back of this page for vocabulary words/expressions.

```
E G X T H I D Y K D A K U I J T O E N V R P B A F
M O G J E K C I E I O V N O W A J R M I E M N R S
E L G N A T C E R N E D L O G C O T Q R U R A A P
T D E N R Z E M Z T B G J E E R Y Y M R E C T Q H
S E O R R B I B Z Y L R Y F L M N P N T T G F M Y
Y N M U O R H V Z H E Y Y T I L D K T A H O N F L
S M E S N Y R T E M M Y S N O I T A L S N A R T L
N E T W S N R E T T A P R E P A P L L A W G A P O
O A R R E A G L I D E R E F L E C T I O N A M L T
I N I N Q J G S N M I E R I H D T S J F N T S N A
T P C R E C U R S I O N P T A U S Q Z C G W V V X
C Q M E L P T M M L L N F N E S C Y R T E M O S I
N R E T T A P E H T F O P U O R G Y R T E M M Y S
U I A T E O E H F V Y K X E C Y I I M I Z O P F R
F G N A P R E S E R V E S T H E P A T T E R N I O
D I D P R O T A T I O N A L S Y M M E T R Y S F T
E D E E T S R E B M U N I C C A N O B I F M H G A
T M T T O E M R L N E T O F E T G I C S O E R O R
A O A T E M W N N O I T R O P O R P E N I V I D E
R T R E Y S A E A F V P N L G X Z C I A O O D U N
E I E S T R I P P A T T E R N R C G T U H M Q L E
T O N O I T A R N E D L O G E I E H H G O E E I G
I N E R F W J L P G I U E Z E A T N M P J F T D E
W F G L J D S P S S P Q E R T M E U G C E E M D E
H E C K A I F S O X T T E X N O I T A L S N A R T
```

1. _____   8. _____

2. _____   9. _____

3. _____   10. _____

4. _____   11. _____

5. _____   12. _____

6. _____   13. _____

7. _____   14. _____

15. _____

16. _____

17. _____

18. _____

19. _____

20. _____

21. _____

22. _____

23. _____

24. _____

25. _____

Chapter 20
Tilings

Chapter Objectives

Check off these skills when you feel that you have mastered them.

- [ ] Calculate the number of degrees in each angle of a given regular polygon.

- [ ] Given the number of degrees in each angle of a regular polygon, determine its number of sides.

- [ ] Define the term tiling (tessellation).

- [ ] List the three regular polygons for which a monohedral tiling exists.

- [ ] When given a mix of regular polygons, determine whether a tiling of these polygons could exist.

- [ ] Explain the difference between a periodic and a nonperiodic tiling.

- [ ] Discuss the importance of the Penrose tiles.

- [ ] Explain why fivefold symmetry in a crystal structure was thought to be impossible.

Guided Reading

Introduction

We examine some traditional and modern ideas about **tiling (tessellation)**, the covering of an area or region of a surface with specified shapes. The beauty and complexity of such designs come from the interesting nature of the shapes themselves, the repetition of those shapes, and the symmetry or asymmetry of arrangement of the shapes.

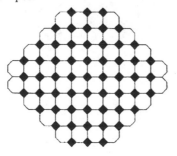

Section 20.1   Tilings with Regular Polygons

⌐ Key idea

A tiling is **monohedral** if all the tiles are the same shape and size; the tiling would consist of repetitions of one figure laid down next to each other.

〰 Example A

Sketch an example of a monohedral tiling using a hexagon.

Solution

The answer is: There are many kinds of hexagonal tilings; here is one example.

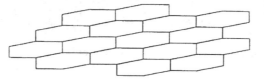

⌐ Key idea

A **tiling** of the plane is a covering of that flat surface with non-overlapping figures.

〰 Example B

a)   Draw a sketch that shows how to tile the plane with squares of two different sizes.

b)   Can you do the same thing with circles?

Solution

a)   The squares must fit together without spaces or overlaps. Here is a way to do it; there are many others.

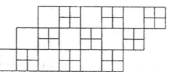

b)   There is no way to fit circles together without overlapping or leaving spaces.

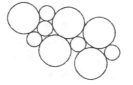

⌐ Key idea

In an **edge-to-edge tiling**, the interior angles at any vertex add up to 360°. An example is the following.

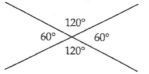

⚬─┬ Key idea

A **regular tiling** uses one tile, which is a regular polygon. Here are only three possible regular tilings.

Triangles   Hexagons   Squares

ᏮᏏ Example C

Explain why a regular edge-to-edge tiling using octagons is impossible.

Solution

A regular tiling using a polygon with n sides is impossible if $n > 6$. This is because the interior angle A, shown below for hexagons and octagons, would be larger than 120° (the angle for a hexagon), so three angles would not fit around a vertex, as it does in the hexagonal tiling.

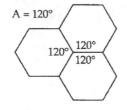

A = 120°

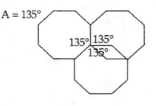

A = 135°

⚬─┬ Key idea

There are eight additional **semi-regular tilings**, using a mix of regular polygons with different numbers of sides.

✎ Question 1

What is the measure of each interior and exterior angle of

a)   a regular quadrilateral?

b)   a regular pentagon?

c)   a regular dodecagon (12 sides)?

Answer

a)   90°; 90°

b)   108°; 72°

c)   150°; 30°

Section 20.2   Tilings with Irregular Polygons

⌖ Key idea

It is easy to adapt the square tiling into a monohedral tiling using a **parallelogram**. Since two triangles together form a parallelogram, any triangle can tile the plane.

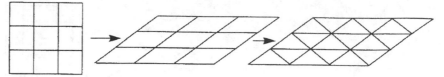

⌖ Key idea

Any **quadrilateral** (four-sided figure), even one that is not convex, can tile the plane. A figure is **convex** if any two points on the figure (including the boundary) can be connected and the line segment formed does not go out of the figure.

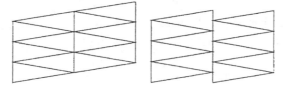

Also, any triangle can tile the plane. A scalene triangle has no two sides the same measure.

✍ Example D

Draw a tiling of the plane with the following figure.

Solution

Here is an edge-to-edge tiling with the given triangle, and another one which is not edge-to-edge.

✍ Example E

Draw a tiling of the plane with the following figure.

Solution

Two copies of the quadrilateral, one of which is rotated by 180°, fit together to form a parallelogram, thus forming an easy tiling.

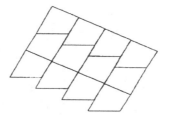

🔑 Key idea

Only certain classes of convex pentagons and hexagons can be used to tile the plane. There are exactly three classes of convex hexagons that can tile a plane. A convex polygon with seven or more sides cannot tile the plane.

🔑 Key idea

The work of artist **M. C. Escher,** famous for his prints of interlocking animals, demonstrates an intimate link between art and mathematics.

Section 20.3   Using Translations

🔑 Key idea

The simplest way to create an Escher-like tiling is through the use of translation. The boundary of each tile must be divisible into matching pairs of opposing parts that interlock.

🔑 Key idea

A single tile can be duplicated and used to tile by **translation** in two directions if certain opposite parts of the edge match each other.

Examples:

These two are based on a parallelogram tiling.

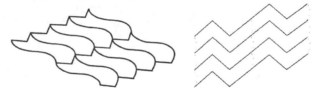

This one is based on a hexagon tiling.

🐌 Example F

a)   Draw a square, replace opposite sides with congruent curved edges, and draw a tiling by translation with the resulting figure.

b)   Can you do the same thing with a triangle?

Solution

The answer is:

a)   Translating horizontally and vertically, opposite sides match up. Here is an illustration of the process for a square-based tile.

b)   You cannot pair up "opposite" sides in a triangle, or any polygon with an odd number of sides. This cannot be done.

Section 20.4   Using Translations Plus Half-Turns

⌐ Key idea

If you replace certain sides of a polygon with matching **centrosymmetric** segments, it may be possible to use the resulting figure to tile the plane by translations and half-turns. The **Conway criterion** can be used to decide if it is possible. The Conway criterion is given on page 770 of your text.

↷ Example G

Start with the following triangle.

Replace the two long sides with centrosymmetric curves and sketch a tiling by translations and half-turns.

Solution

The answer is: This shows the process and the resulting tiling:

Here are the new sides.   This is the tile.

Here is the tiling.

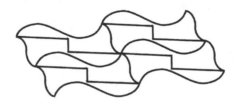

⌐ Key idea

Many fascinating and beautiful examples of these principles are found in the designs of the renowned graphic artist M. C. Escher.

⌐ Key idea

Periodic tilings have a fundamental region that is repeated by translation at regular intervals.

⌐ Key idea

A **fundamental region** consists of a tile, or block of tiles, with which you can tile a plane using translations at regular intervals.

✎ Question 2

Are there two fundamental regions for this tiling?

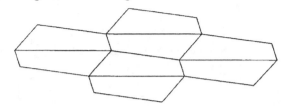

Answer

yes

Section 20.5   Nonperiodic Tilings

⌇ Key idea

A tiling may be **nonperiodic** because the shape of the tiles varies, or the repetition of the pattern by translation varies. The Penrose tiles are an important example of a set of two tiles which can be used only to tile the plane nonperiodically.

⌇ Key idea

Penrose tilings exhibit the following properties: self-similarity (inflation and deflation), the golden ratio (quasiperiodic repetition in that proportion), and partial five-fold rotational symmetry.

⌇ Key idea

Applying principles of tiling to three dimensional crystals, **Barlow's law** states that a crystal cannot have more than one center of fivefold rotational symmetry.

⌇ Example I

What chemical crystal exhibits strict fivefold symmetry?

Solution

Crystals are periodic three-dimensional objects, and it follows from Barlow's law that truly periodic patterns can only have twofold, threefold, fourfold, or sixfold symmetry. The answer is none; certain quasicrystals exhibit limited **fivefold symmetry**.

Homework Help

Exercises 1 – 10
Carefully read Section 20.1 before responding to these exercises.

Exercises 11 – 12
Carefully read Section 20.2 before responding to these exercises.

Exercises 13 – 26
Carefully read Section 20.3 before responding to these exercises.

Exercises 27 – 38
Carefully read Section 20.4 before responding to these exercises.

Exercises 39 – 50
Carefully read Section 20.5 before responding to these exercises.

Do You Know the Terms?

Cut out the following 24 flashcards to test yourself on Review Vocabulary. You can also find these flashcards at http://www.whfreeman.com/fapp7e.

Chapter 20 Tilings **Barlow's law, or the crystallographic restriction**	Chapter 20 Tilings **Centrosymmetric**
Chapter 20 Tilings **Convex**	Chapter 20 Tilings **Conway criterion**
Chapter 20 Tilings **Edge-to-edge tiling**	Chapter 20 Tilings **Equilateral triangle**
Chapter 20 Tilings **Exterior angle**	Chapter 20 Tilings **Fundamental region**

Symmetric by 180° rotation around its center.	A law of crystallography that states that a crystal may have only rotational symmetries that are twofold, threefold, fourfold, or sixfold.
A criterion for determining whether a shape can tile by means of translations and half-turns.	A geometric figure is this if for any two points on the figure (including its boundary), all the points on the line segment joining them also belong to the figure (including its boundary).
A triangle with all three sides equal.	A tiling in which adjacent tiles meet only along full edges of each tile.
A tile or group of adjacent tiles that can tile by translation.	The angle outside a polygon formed by one side and the extension of an adjacent side.

Chapter 20
Tilings

Interior angle

Chapter 20
Tilings

Monohedral tiling

Chapter 20
Tilings

***n*-gon**

Chapter 20
Tilings

Nonperiodic tiling

Chapter 20
Tilings

Parallelogram

Chapter 20
Tilings

Par-hexagon

Chapter 20
Tilings

Periodic tiling

Chapter 20
Tilings

Quadrilateral

A tiling with only one size and shape of tile (the tile is allowed to occur also in "turned-over," or mirror-image, form).

The angle inside a polygon formed by two adjacent sides.

A tiling in which there is no repetition of the pattern by translation.

A polygon with n sides.

A hexagon whose opposite sides are equal and parallel.

A convex quadrilateral whose opposite sides are equal and parallel.

A polygon with four sides.

A tiling that repeats at fixed intervals in two different directions, possibly horizontal and vertical.

Chapter 20 Tilings **Regular polygon**	Chapter 20 Tilings **Regular tiling**
Chapter 20 Tilings **Rhombus**	Chapter 20 Tilings **Scalene triangle**
Chapter 20 Tilings **Semiregular tiling**	Chapter 20 Tilings **Tiling**
Chapter 20 Tilings **Translation**	Chapter 20 Tilings **Vertex figure**

A tiling by regular polygons, all of which have the same number of sides and are the same size; also, at each vertex, the same kinds of polygons must meet in the same order.

A polygon all of whose sides and angles are equal.

A triangle no two sides of which are equal.

A parallelogram all of whose sides are equal.

A covering of the plane without gaps or overlaps.

A tiling by regular polygons; all polygons with the same number of sides must be the same size.

The pattern of polygons surrounding a vertex in a tiling.

A rigid motion that moves everything a certain distance in one direction.

Practice Quiz

1. What is the measure of the exterior angle of a regular decagon?

    a. 36°

    b. 72°

    c. 144°

2. Which of the following polygons will tile the plane?

    a. regular pentagon

    b. scalene triangle

    c. regular octagon

3. A semi-regular tiling is made with two dodecagons and another polygon at each vertex. Use measures of interior angles to determine which other polygon is required.

    a. triangle

    b. square

    c. hexagon

4. Squares and triangles can be used to form a semi-regular tiling of the plane. How many of each figure is needed?

    a. 2 triangles, 3 squares

    b. 4 triangles, 1 square

    c. 3 triangles, 2 squares

5. Are Penrose Tilings non-periodic?

    a. not enough information

    b. no

    c. yes

6. Which of the following is true?

    I: Only convex polygons will tile the plane.

    II: Any quadrilateral will tile the plane.

    a. Only I is true.

    b. Only II is true.

    c. Both I and II are true.

7. Which of the following is true?

    I. It is possible to tile the plane with a square using translations.

    II. It is possible to tile the plane with a square using half turns.

    a. I only

    b. II only

    c. both I and II

8.   Can the tile below be used to tile the plane?

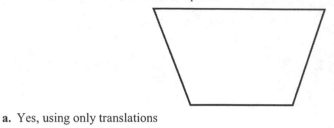

   a. Yes, using only translations
   b. Yes, using translations and half turns
   c. No

9.   The Penrose tilings use two different figures with how many sides?
   a. 3 and 5
   b. 4 and 4
   c. 5 and 5

10. Which polygon can be altered to form an Escher-type tiling by translation only?
   a. equilateral triangle
   b. any convex quadrilateral
   c. regular hexagon

Word Search

Refer to pages 783 – 784 of your text to obtain the Review Vocabulary. There are 25 hidden vocabulary words/expressions in the word search below. Both *Barlow's law and Crystallographic restriction* appear in the word search. Also, *Periodic tiling* appears separately from *Nonperiodic tiling,* as do *Regular tiling* and *Semiregular tiling.* It should be noted that spaces and hyphens are removed as well as apostrophes. There is additional space on the back of this page for vocabulary words/expressions.

```
C S T S X B E B N A G F R L I G N I L I T R A L U G E R
C A U E R L L Y T N R N J K R N E E S M O H T W D X R J
N E P X G D S Q O K N N M A Z K N C T X I K N H U A E O
O I N T E R I O R A N G L E E R U G I F X E T R E V L P
I G E T R A N S L A T I O N R G Y H R G A N E Q F G D P
G N M D R E N E E T S W R P E M E E D O T G O E N E M E
E I D N Z O W E P T G Q D O L O U J K E U M D M R P V R
R L O W J N S C A L E N E T R I A N G L E G U L F M P I
L I N E N O G Y L O P R A L U G E R A B E W S E H E E O
A T E M E R I S M U C Q O X M L J R D T R Y D Y E I S D
T L A Q L G G J U M T E W L Y H T N O G A X E H R A P I
N A T S U S R R T Q E O W G A I Q E F J H E X E V N O C
E R M M R I S S K N A T T S L R D Q M O N F R R E E E T
M D D I H N L P S C M Q R I R G E I R B E I F S O E A I
A E L Q G N M A N P P S N I E O N T E N E R E I E P H L
D H I V T A S E T E L G T T C O F G A B Y C Q E O E T I
N O I T C I R T S E R C I H P A R G O L L A T S Y R C N
U N T H N O E R N E R L W Q T E G E P N I U I N A E P G
F O N S L X C I R C I A Y E L G N A R O I R E T X E K S
F M O C H T L S M N C M L J M K R D G A E S D S P H W S
O N T S I R U T G N I L I T X N K N E L C Z M A M E T P
P D G N I L I T R A L U G E R I M E S T S D N S U I A S
T A N T M F F A M W C M D I Y I P L E E E U Y S E Q W E
F X P O G W R T C E N E Q S I W A E M W R E B A Q D I R
B S Z G N I L I T C I D O I R E P N O N D B E M N F L W
C L D E N M E S R N P R F D A E A N G T A K R Y O S D R
B A R L O W S L A W E M A R G O L E L L A R A P O H F P
J F I I T N K D A B C O N W A Y C R I T E R I O N L R S
```

1. _____   7. _____

2. _____   8. _____

3. _____   9. _____

4. _____   10. _____

5. _____   11. _____

6. _____   12. _____

13. _____   20. _____

14. _____   21. _____

15. _____   22. _____

16. _____   23. _____

17. _____   24. _____

18. _____   25. _____

19. _____

Chapter 21
Savings Models

Chapter Objectives

Check off these skills when you feel that you have mastered them.

☐ Apply the simple interest formula to calculate the balance of a savings account.

☐ Apply the compound interest formula to calculate the balance of a savings account.

☐ Describe the difference between arithmetic and geometric growth.

☐ Calculate the APY for a compound interest account.

☐ Apply the interest formula for continuous compounding to calculate the balance of a savings account.

☐ Find the sum of a geometric series.

☐ Use the savings formula to determine required deposits into a sinking fund.

☐ Calculate depreciation of a financial asset, given a negative growth rate.

☐ Use the Consumer Price Index to find the current cost of goods.

Guided Reading

Introduction

We investigate and try to model situations of growth in economics and finance, examining the increase or decrease in value of investments and various economic assets. These models will apply similarly to the growth of biological populations, money in a bank account, pollution levels, and so on. In the same vein, managing a financial entity like a trust fund is similar to managing a renewable biological resource.

Section 21.1 Arithmetic Growth and Simple Interest

⚬⊸ **Key idea**

The initial balance of an account, such as a savings account, is the **principal**. At the end of a fixed amount of time, **interest** is added to the account.

⚬⊸ **Key idea**

If a population, measured at regular time intervals, is experiencing **arithmetic growth** (also called **linear growth**), then it is gaining (or losing) a constant amount with each measurement.

↷ Example A

If an account has an initial value of $500 and gains $150 at each interval, write the sequence of population values for the first ten intervals.

Solution

Starting with the initial value $500, successively add $150. The result is: $500, 650, 800, 950, 1100, 1250, 1400, 1550, 1700, 1850.

⚷ Key idea

A financial account which is paying **simple interest** will grow arithmetically in value. If we let I be the interest earned, P be the principal, r be the annual rate of interest, A be the total amount (including principal and interest), and t be the number of years, the formulas involving simple interest are as follows.

$$I = Prt \text{ and } A = P + I = P + Prt = P(1 + rt)$$

✎ Question 1

If $5000 is invested at an annual rate of 2.4% with simple interest, how much money is in the account after 4 years?

Answer

$5480

Section 21.2   Geometric Growth and Compound Interest

⚷ Key idea

A savings account which earns **compound interest** is growing geometrically. At the end of the first year, the initial balance, or principal, is increased by the interest payment. Each successive year, the new balance is the previous balance plus interest, paid as a fixed percentage of that balance. The value of the savings account is determined by the **principal** (or **initial balance**), the **rate of interest**, and the **compounding period**.

↷ Example B

Suppose you deposit a $2000 principal to start up a bank account with an annual interest rate of 8%, compounded quarterly. How much money will you have in the account at the end of the first year?

Solution

Each quarter the value of the account grows by 1/4 of the annual interest of 8%. This means the account balance is multiplied by 1.02 four times in the course of the year. Here is a table of account values.

initial deposit	quarter 1	quarter 2	quarter 3	quarter 4
$2000	$2040	$2080.80	$2122.42	$2164.86

There will be $2164.86 at the end of the year.

⚷ Key idea

The annual interest rate of a savings account is called the **nominal rate**. With compounding, the actual realized percentage is higher; it is called the **effective annual rate**, or **EAR**.

⚷ Key idea

The nominal rate i for a period during which no compounding is done is given by $i = \dfrac{r}{n}$, where r is the nominal annual rate and n is the number of times interest is compounded per year.

↺ Example C

For an account earning 9% interest compounded monthly, what is the nominal rate i for a one-month period?

Solution

Because $r = 9\%$ and $n = 12$, $i = \frac{0.09}{12} = 0.0075$ or 0.75% per month.

⚷ Key idea

The **compound interest formula** for the value of a savings account after compounding periods is as follows.

$$A = P(1+i)^n$$

Here, P is the principal and i is the interest rate per compounding period.

↺ Example D

If $1000 is deposited in an account earning 12% interest compounded annually, what will be the value of the account:

a)   after 5 years?

b)   after 20 years?

Solution

a)   With annual compounding, we have one interest payment each year, or five in five years. Because $i = 12\% = 0.12$, the formula for the value of the account after 5 years is

$A = 1000(1+0.12)^5 = \$1762.34$.

b)   After 20 years we have $A = 1000(1+0.12)^{20} = \$9646.29$.

✎ Question 2

If $5000 is deposited in an account earning 2.4% interest compounded quarterly, what will be the value of the account:

a)   after 5 years?

b)   after 20 years?

Answer

a) $5635.46

b) $8068.79

↺ Example E

If $1000 is deposited in a different account earning 12% interest compounded monthly, what will be the value of the account:

a)   after 5 years?

b)   after 20 years?

Solution

a)   With monthly compounding, we have 12 interest payments per year, or 60 in five years. Now the formula for the value of the account after 5 years is $A = 1000\left(1+\frac{0.12}{12}\right)^{60} = \$1816.70$.

b)   After 20 years, we have 240 interest periods, so $A = 1000\left(1+\frac{0.12}{12}\right)^{240} = \$10,892.55$.

⮡ Key idea

The present value P of an amount A to be paid in the future, after earning compound interest or n compounding periods at a rate i per compound period is as follows.

$$P = \frac{A}{(1+i)^n}$$

⮡ Key idea

If a population is experiencing **geometric (exponential) growth**, then it is increasing or decreasing by a fixed proportion of its current value with each measurement. The proportion is called the growth rate of the population.

⮡ Key idea

Accounts earning compound interest will grow more rapidly than accounts earning simple interest. In general, geometric growth (such as compound interest) is much more dramatic than arithmetic growth (such as simple interest).

Section 21.3   A Limit to Compounding

⮡ Key idea

For a nominal interest rate r compounded n times per year, **the annual effective interest rate**, or **APY**, is $\left(1+\frac{r}{n}\right)^n - 1$.

🖉 Example F

What is the APY for 7.5% compounded monthly?

Solution

Because $r = 0.075$ and $n = 12$, the APY is $\left(1+\frac{r}{n}\right)^n - 1 = \left(1+\frac{0.075}{12}\right)^{12} - 1 = (1.00625)^{12} - 1 = 0.0776$ or 7.76%.

⮡ Key idea

Compounding interest more often results in a higher value in the account because interest is earned earlier and is included with the principal in the next compounding period. However, the more often interest is compounded, the less significant this increase becomes. The limit is reached when interest is **compounded continuously**. The formula for finding the account balance when interest is compounded continuously is $A = Pe^{rt}$, where r is the nominal interest rate and t is the time in years. e is a constant which is approximately equal to 2.718281828. The pattern of decimals do not repeat and does not yield a rational number.

🖉 Example G

If $1000 is deposited into an account earning 9% interest compounded continuously, find the value of the account after

a)   5 years.

b)   10 years.

Solution

a)   Using a calculator we find that $A = 1000e^{(0.09)(5)} = 1000e^{0.45} = \$1568.31$.

b)   Similar to what was done in part a), we have $A = 1000e^{(0.09)(10)} = 1000e^{0.90} = \$2459.60$.

⌐ Key idea

There is virtually no difference whether a bank treats a year as 365 days or 360 days. The **365 over 365 method** with a daily nominal rate of $\frac{r}{365}$ is the usual method for daily compounding. The **360 over 360 method** with a daily nominal rate of $\frac{r}{360}$ is the usual method for loans with equal monthly installments.

Note: In the Student Solutions Guide, the solution to Exercise 13 should use the formulas $P\left(1+\frac{r}{360}\right)^{360}$ and $P\left(1+\frac{r}{365}\right)^{365}$ in parts b and c, respectively.

✐ Question 3

If $10,000 is deposited into an account with annual rate of 2.7%, what is the amount in the account after 1 year if it is compounded

a) continuously?
b) using the 365 over 365 method?
c) using the 360 over 360 method?

Answer
a) $10,273.68
b) $10,273.67
c) $10,273.67

Section 21.4   A Model for Investment

⌐ Key idea

A **geometric series** with first term 1, common ratio x, and n terms is the following.
$$1+x+x^2+...+x^{n-1}$$
The sum of these terms is $\dfrac{x^n-1}{x-1}$.

⌐ Key idea

We can accumulate a desired amount of money in a savings account by a fixed date by making regular deposits at regular intervals – a **sinking fund**. With a uniform deposit of d dollars at the end of each interval, and an interest rate of i per interval, the **savings formula** predicts that the value of the account after n intervals will be as follows.

$$A = d\left[\frac{(1+i)^n-1}{i}\right]$$

This formula is obtained by summing a geometric series of accumulated deposits and interest.

✍ Example H

Rosetta's flower shop is growing, and she will need $6000 five years from now to pay for an addition to her greenhouse. How much would she have to deposit each month in a sinking fund with a 6% annual interest rate to accumulate this amount?

Solution

Here, d is the unknown amount to be deposited, $i = \frac{0.06}{12} = 0.005$ is the monthly interest rate and $n = 5 \times 12 = 60$, the number of deposits in 5 years. Then we can solve for d in the savings formula as follows.

$$A = d\left[\frac{(1+i)^n - 1}{i}\right]$$

$$6000 = d\left[\frac{(1+0.005)^{60} - 1}{0.005}\right]$$

$$6000 = d(69.770030510)$$

$$d = \frac{6000}{69.770030510} = \$86.00$$

Rosetta would deposit $86 at the end of each month for 5 years.

☞ Key idea

An **annuity** pays a specified number of equal payments at equally spaced time intervals. A sinking fund is the reverse of an annuity. With a sinking fund you make payments and with an annuity you receive payments.

Section 21.5   Exponential Decay and the Consumer Price Index

☞ Key idea

Geometric growth with a negative growth rate is called **exponential decay**. Examples are depreciation of the value of a car and decay in the level of radioactivity of a given quantity of a radioactive isotope. The quantity is declining at a rate which is negative and proportional to its size; the proportion l is called the decay constant.

✍ Example I

If a car costs $12,000 new and its value depreciates at 20% per year, give a formula for its value after n years and predict its value in five years.

Solution

The general formula is $A = P(1+i)^n$, where $P = \$12,000$, $i = -0.2, n = 5$. The formula is $A = 12,000(0.8)^n$ and the value $A = \$3932.16$ can be found using a calculator.

☞ Key idea

With inflation, the value of currency declines. If the rate of inflation is a, then the present value of a dollar in one year is given by the formula $\dfrac{\$1}{1+a} = \$1 - \dfrac{\$a}{1+a}$.

🖋 Example J

Assuming constant 7% inflation, what would the present value of a $50 bill be in five years?

Solution

The present value of one dollar after one year is given by $\dfrac{\$1}{1+0.07} = 0.93458$. or 93.46 cents. After five years, the present value has shrunk by this factor five times. Thus, after five years one dollar has a value of $\left(\dfrac{\$1}{1+0.07}\right)^5 = 0.7130$, or 71.3 cents. Multiplying this by 50 we have $35.65.

⌐→ Key idea

The **Consumer Price Index (CPI)** is the measure of inflation. The CPI compares the current cost of certain goods, including food, housing, and transportation, with the cost of the same (or comparable) goods in a base period.

⌐→ Key idea

To convert the cost of an item in dollars for one year to dollars in a different year, use the following proportion.

$$\frac{\text{Cost in year A}}{\text{Cost in year B}} = \frac{\text{CPI for year A}}{\text{CPI for year B}}$$

You will need to use Table 21.5 on page 816 of your text to obtain the CPI for a particular year.

🖋 Example K

Rodney bought a house in 1967 for $25,000 and sold it in 2004. How much would the house be worth in 2004 dollars?

Solution

Set up the following proportion using values from Table 21.5.

$$\frac{\text{Cost in 1967}}{\text{Cost in 2004}} = \frac{\text{CPI for year 1967}}{\text{CPI for year 2004}}$$

$$\frac{\$25,000}{\text{Cost in year 2004}} = \frac{33.4}{189.5}$$

$$\$25,000(189.5) = 33.4(\text{Cost in year 2004})$$

$$\text{Cost in year 2004} = \frac{\$4,737,500}{33.4} = \$141,841.3174$$

The house is worth around $142,000.

Section 21.6   Real Growth and Valuing Investments

⌐→ Key idea

An investment is affected by the rate of inflation. An investment that grows at, say, 6.5% per year will not actually gain purchasing power at 6.5% per year if inflation is considered.

⌐→ Key idea

If an investment grows at an annual rate r and the rate of inflation is a, the real growth rate g is given by the following.

$$g = \frac{r-a}{1+a}$$

↶ Example L

In late 2001 the inflation rate was about 2.9%. If you invested in a savings account with an annual interest rate of 6.5%, what was the real growth rate of this investment?

Solution

In this case, $r = 6.5\% = 0.065$ and $a = 2.9\% = 0.029$, so we have the following.

$$g = \frac{r-a}{1+a} = \frac{0.065-0.029}{1+0.029} = \frac{0.036}{1.029} = 0.03499$$

☐→ Key idea

Buyers of stock in a company receive part of the profits of the company in the form of cash payments, known as **dividends.**

☐→ Key idea

An **overvalued** stock is one that has a current price that is too high compared to the company's expected earnings.

☐→ Key idea

Modeling the value of a stock requires considering the growth rate of the annual dividend as well as the **discount rate**. The discount rate for a stock takes into account the effects of inflation, the level of risk for that particular stock, and the value of interest on alternate investments.

☐→ Key idea

Suppose you want to buy a stock at the beginning of the year and D was last year's dividends, paid at the beginning of this year to those that owned stock on a particular date, g is the annual rate the company expands and prospers and r is the discount rate per year, then the present value of next year's dividends is $D\dfrac{1+g}{1+r}$.

☐→ Key idea

An option to buy a stock at a certain price by a certain time is an example of a "financial derivative." The true value of a derivative depends on the current value and the probabilities that the stock will go up or down within the option's time frame. The famous "**Black-Scholes formula**" is often used to value financial derivatives.

Homework Help

Exercises 1 – 10
Carefully read Sections 21.1 and 21.2 before responding to these exercises.

Exercises 11 – 16
Carefully read Section 21.3 before responding to these exercises.

Exercises 17 – 28
Carefully read Section 21.4 before responding to these exercises.

Exercises 29 – 42
Carefully read Section 21.5 before responding to these exercises.

Exercises 43 – 50
Carefully read Section 21.6 before responding to these exercises.

Do You Know the Terms?

Cut out the following 23 flashcards to test yourself on Review Vocabulary. You can also find these flashcards at http://www.whfreeman.com/fapp7e.

Chapter 21 Savings Models **Annual percentage yield (APY)**	Chapter 21 Savings Models **Annuity**
Chapter 21 Savings Models **Arithmetic growth**	Chapter 21 Savings Models **Compound interest**
Chapter 21 Savings Models **Compound interest formula**	Chapter 21 Savings Models **Compounding period**
Chapter 21 Savings Models **Constant dollars**	Chapter 21 Savings Models **Continuous compounding**

A specified number of payments at equal intervals of time.	The effective interest rate per year.
Interest that is paid on both the original principal and accumulated interest.	Growth by a constant amount in each time period.
The fundamental interval for compounding, within which no compounding is done.	Formula for the amount in an account that pays compound interest periodically. For an initial principal P and effective rate i per compounding period, the amount after n compounding periods is $A = P(1+i)^n$.
Payment of interest in an amount toward which compound interest tends with more and more frequent compounding.	Costs are expressed in constant dollars if inflation or deflation has been taken into account by converting the costs to their equivalent in dollars of a particular year.

Chapter 21 Savings Models **Current dollars**	Chapter 21 Savings Models *e*
Chapter 21 Savings Models **Effective annual rate (EAR)**	Chapter 21 Savings Models **Effective rate**
Chapter 21 Savings Models **Exponential decay**	Chapter 21 Savings Models **Exponential growth**
Chapter 21 Savings Models **Geometric growth**	Chapter 21 Savings Models **Geometric series**

The base for continuous compounding, geometric (exponential) growth, and natural logarithms.

The actual cost of an item at a point in time; inflation or deflation before or since then has not been taken into account.

The actual percentage rate, taking into account compounding. It equals the rate of simple interest that would realize exactly as much interest over the same period of time.

The effective rate per year.

Geometric growth.

Geometric growth at a negative rate.

A sum of terms, each of which is the same constant times the previous term; that is, the terms grow geometrically.

Growth proportional to the amount present.

Chapter 21 Savings Models **Interest**	**Chapter 21** Savings Models **Linear growth**
Chapter 20 Tilings **Nominal rate**	**Chapter 21** Savings Models **Present value**
Chapter 21 Savings Models **Principal**	**Chapter 21** Savings Models **Simple interest**
Chapter 21 Savings Models **Sinking fund**	

Arithmetic growth.	Money earned on a savings account or a loan.
The value today of an amount to be paid or received at a specific time in the future, as determined from a given interest rate and compounding period.	A stated rate of interest for a specified length of time; a nominal rate does not take into account any compounding.
The method of paying interest only on the initial balance in an account, not on any accrued interest.	Initial balance.
	A savings plan to accumulate a fixed sum by a particular date.

Practice Quiz

1. If you deposit $1000 at 5.2% simple interest, what is the balance after three years?
    a. $1140.61
    b. $1164.25
    c. $1156.00

2. Suppose you invest $250 in an account that pays 2.4% compounded quarterly. After 30 months, how much is in your account?
    a. $265.41
    b. $299.14
    c. $268.00

3. A home was purchased in 1976 for $25,000. How much would a comparable house be worth in 1999? (Assume the 1976 CPI is 56.9 and the 1999 CPI is 166.2.)
    a. about $25,290
    b. about $73,023
    c. about $85,589

4. What is the APY for 4.8% compounded monthly?
    a. 4.8%
    b. 4.9%
    c. 5.2%

5. If a bond matures in 3 years and will pay $5000 at that time, what is the fair value of it today, assuming the bond has an interest rate of 4.5% compounded annually?
    a. $4381.48
    b. $4325.00
    c. $4578.65

6. The stated rate of interest for a specific length of time is called the:
    a. APY
    b. effective rate
    c. nominal rate

7. Suppose your stock portfolio increased in value from $12,564 to $12,870 during a one-month period. What is the APY for your gains during this period?
    a. 29.2%
    b. 32.6%
    c. 33.5%

8. Suppose you buy a new tractor for $70,000. It depreciates steadily at 8% per year. When will it be worth approximately $10,000?
    a. after about 11 years
    b. after about 18 years
    c. after about 25 years

9. Suppose a stock worth $50 today is equally likely to be worth $100 or $25 a year from today. What is a fair price to pay today for the option to buy this stock a year from now?

    a. about $7.50

    b. about $10

    c. about $12.50

10. Suppose a stock worth $50 today is equally likely to be worth $75 or $35 a year from today. What is the expected value of this stock a year from now?

    a. $75

    b. $55

    c. $50

Word Search

Refer to pages 822 – 823 of your text to obtain the Review Vocabulary. There are 22 hidden vocabulary words/expressions in the word search below. It should be noted that no abbreviations appear and spaces are removed. *Interest*, *Compound Interest*, and *Compound Interest Formula*, appear separately. Also, the symbol *e* is not considered part of the search.

```
R A A Z B S I M P L E I N T E R E S T H E E L S T
T R N S D Y E N E W X U R B D E W H E I S M Q E S
S I N T E R E S T F P K A N U U A P T I H N N S E
B T U V W E K A E R O P F H S L E Z A E T O T G R
N H A I E I X P O N E N T I A L G R O W T H I E
I M L Z X P A R M C E F T M N V G N L D O F A O T
O E P D E H C B E O N F P M K T P I A I R Y E E N
O T E U T P R J T N T E E A I N E D N B G E A W I
S I R A T S H F A S I C R R N E R N I R R P A N D
R C C Y U S A I R T A T D D G S I U M G A E H F N
K G E L I R P D L A L I G W F E O O O W E E E E U
P R N P A A N B A N D V T T U R D P N K N Z Z S O
Q O T O U L E K U T E E C B N P E M V J I R R G P
I W A H R L H C N D C R F S D G A O W P L N N R M
I T G I N O M I N O A A P L A P I C N I R P R E O
L H E M K D S E A L Y T I U N N A S J O U S N A C
E S Y C B T N H E L E E A R N E G U T N C C J P T
P E I D M N O C V A D H D F S T C O A O I O T A S
R A E N F E D O I R E P G N I D N U O P M O C N G
D A L U M R O F T S E R E T N I D N U O P M O C C
O C D O S R I E C J H T W O R G C I R T E M O E G
S F G S T U I E E G J O R L I F F T G B X Y U G R
S D E W O C S S F I Y D D K G K S N Z I M P J E U
Q R Q N T F A T F G R W D E X I S O R B O P I D Y
N A C N E E D A E G E O M E T R I C S E R I E S E
```

1. _____ 12. _____
2. _____ 13. _____
3. _____ 14. _____
4. _____ 15. _____
5. _____ 16. _____
6. _____ 17. _____
7. _____ 18. _____
8. _____ 19. _____
9. _____ 20. _____
10. _____ 21. _____
11. _____ 22. _____

Chapter 22
Borrowing Models

Chapter Objectives

Check off these skills when you feel that you have mastered them.

☐ Know the basic loan terms principal and interest.

☐ Be able to solve the simple interest formula to find the amount of a loan over time.

☐ Know the difference between a discounted loan and an add-on loan.

☐ Understand the compound interest formula and use it to find the amount of a loan over time.

☐ Use loan terminology to explain the difference between the nominal rate, effective rate, effective annual rate (EAR), and the annual percentage rate (APR).

☐ Use the amortization formula to determine the payments required to fully amortize a loan.

☐ Find the APR and EAR for a loan.

☐ Understand how an annuity functions and be able to use the annuity formula.

Guided Reading

Introduction

Financial institutions would not be able to offer interest-bearing accounts such as savings accounts unless they had a way to make money on them. To do this, financial institutions use the money from savings accounts to make loans. The loans provide interest for the loaning institution and also enable individuals to make major purchases such as buying a house or a car. Using a credit card is also a way of taking out a loan. This chapter looks at some of the common types of loans.

Section 22.1   Simple Interest

⌗ Key idea
The initial amount borrowed for a loan is the **principal**. **Interest** is the money charged on the loan, based on the amount of the principal and the type of interest charged.

⌗ Key idea
Simple interest uses a fixed amount of interest, which is added to the account for each period of the loan. The amount of interest owed after t years for a loan with principal P and annual rate of interest r is given by $I = Prt$. The total amount A of the loan after this time is $A = P(1 + rt)$.

∿ Example A

A bank offers a $2500 loan and charges 6% simple interest. How much interest will be charged after four years? What is the amount of the loan after four years?

Solution

Because $P = \$2500$, $r = 0.06$, and $t = 4$ years, we have $I = Prt = \$2500(0.06)(4) = \$600$. The amount of the loan after 4 years is $\$2500 + \$600 = \$3100$.

⊶ Key idea

An **add-on loan** is borrowed at an amount P that is to be repaid with n payments in t years. The interest is simple interest at an annual rate of r. Since the total amount of the loan is $A = P(1 + rt)$, each payment, d, will be $d = \dfrac{P(1 + rt)}{n}$.

∿ Example B

A 7% add-on loan is to be repaid in monthly installments over 10 years. The total amount borrowed was $3000. How much is the monthly payment?

Solution

This loan will have 120 payments, because it is paid 12 times per year for 10 years. The amount of the loan will be $A = \$3000(1 + 0.07(10)) = \$3000(1.7) = \$5100$. Dividing this by the 120 payments, we have a monthly payment of $d = \dfrac{\$5100}{120} = \$42.50$.

✐ Question 1

A 9% add-on loan is to be repaid in monthly installments over 14 years. The total amount borrowed was $4500. How much is the monthly payment?

Answer

$60.54

⊶ Key idea

A discounted loan is borrowed at an amount P that is to be repaid in t years. The interest is simple interest at an annual rate of r, just as with an add-on loan. However, the interest is subtracted from the amount given to the borrower at the time the loan is made. So the borrower only gets $P - I = P - Prt = P(1 - rt)$, but still needs to pay back the principal P.

∿ Example C

A 4% discounted loan is to be repaid in monthly installments over 10 years. The total amount borrowed was $3000.

a)   How much is the monthly payment?

b)   How much money does the borrower actually get at the beginning of the loan?

Solution

a)   The total amount to be paid over 10 years is $P = \$3000$. Because it is being paid 12 times per year, the monthly payment is $\dfrac{\$3000}{120} = \$25.00$.

b)   The amount the borrower actually gets at the beginning of the loan is the following.

$$P(1 - rt) = \$3000(1 - (0.04)(10)) = \$3000(0.6) = \$1800.00$$

🖉 Question 2

A 2.5% discounted loan is to be repaid in monthly installments over 5 years. The total amount borrowed was $6000.

a) How much is the monthly payment?

b) How much money does the borrower actually get at the beginning of the loan?

Answer

a) $100

b) $5250

↩ Example D

How much would the discount-loan borrower in Example C need to borrow to get $3000 at the start of the loan?

Solution

Call the loan amount x. The borrower will receive $x(1-rt) = x(1-(0.04)(10)) = 0.6x$. We want this to be equal to $3000, so solve $0.6x = \$3000$ to get $x = \dfrac{\$3000}{0.6} = \$5000$.

Section 22.2   Compound Interest

☞ Key idea

When interest is compounded, the interest that is charged is added to the principal. The next interest calculation is based on the new amount, so the interest from the previous period is now earning interest as well.

☞ Key idea

If a loan has a principal P with an interest rate of i per compounding period, then the amount owed on the loan after n compounding periods is $A = P(1 + i)^n$. This is assuming no repayments.

↩ Example E

Max borrows $500 at 3% interest per month, compounded monthly. If he pays the loan back in two years, how much will he owe?

Solution

Because the interest is compounded monthly for two years, $n = 12 \times 2 = 24$. So the amount owed is $A = P(1 + i)^n = \$500(1 + 0.03)^{24} = \$1016.40$.

☞ Key idea

The **nominal rate** for a loan is the stated interest rate for a particular length of time. The nominal rate does not take compounding into account.

☞ Key idea

The effective rate for a loan does reflect compounding. The **effective rate** is the actual percentage that the loan amount increases over a length of time.

☞ Key idea

When the effective rate is given as a rate per year, it is called the **effective annual rate** (**EAR**).

⌐ Key idea

The nominal rate for a length of time during which no compounding occurs is denoted as i. The **annual percentage rate (APR)** is given by the rate of interest per compounding period, i, times the number of compounding periods per year, n. Thus, $\text{APR} = i \times n$.

⌐ Example F

A credit card bill shows a balance due of $750 with a minimum payment of $15 and a monthly interest rate of 1.62%. What is the APR?

Solution

Because the monthly interest rate is $i = 0.0162$ and the account is compounded monthly, the APR is $i \times n = (0.0162)(12) = 0.1944$,   The balance due and minimum payment information is not needed to solve this.

Section 22.3   Conventional Loans

⌐ Key idea

When the regular amounts d are payments on a loan, they are said to **amortize** the loan. The **amortization formula** equates the accumulation in the savings formula with the accumulation in a savings account given by the compound interest formula. This is a model for saving money to pay off a loan all at once, at the end of the loan period.

⌐ Key idea

For a loan of A dollars requiring n payments of d dollars each, and with interest compounded at rate i in each period, the amortization formula is as follows.

$$A = d\left[\frac{1-(1+i)^{-n}}{i}\right] \text{ or } d = \frac{Ai}{1-(1+i)^{-n}}$$

⌐ Example G

David takes out a conventional loan to purchase a car. The interest rate is 4.8% compounded quarterly and David has five years to repay the $8000 he borrowed. What are David's monthly payments?

Solution

In this case, A = $8000. Because there are four compounding periods for each of the five years, $n = 4 \times 5 = 20$.   To find the interest rate per compounding period, divide by four because there are four compounding periods per year: $i = \frac{0.048}{4} = 0.012$.   So given $d = \frac{Ai}{1-(1+i)^{-n}}$ we have the following.

$$d = \frac{\$8000(0.012)}{1-(1+0.012)^{-20}} = \frac{\$96}{0.2122475729} = \$452.30$$

✎ Question 3

Ivan takes out a conventional five-year loan to purchase a $20,000 car. The interest rate is 3.6% compounded monthly. What are Ivan's monthly payments?

Answer

$364.73 (with pure rounding) or $364.74 (with rounding up)

⌖ Key idea

The **effective annual rate (EAR)** takes into account monthly compounding. For a loan with n compounding periods per year, with interest compounded at rate i in each compounding period, the EAR can be found using $\text{EAR} = (1 + i)^n - 1$.

ᒉᔆ Example H

A credit card bill shows a balance due of $750 with a minimum payment of $15 and a monthly interest rate of 1.62%. What is the EAR?

Solution

Because the monthly interest rate is $i = 0.0162$ and the account is compounded monthly, the EAR is as follows.

$$(1 + 0.0162)^{12} - 1 = 0.2127, \text{ or } 21.27\%$$

The balance due and minimum payment information is not needed to solve this.

⌖ Key idea

After owning a home for a period of time, one builds **equity** in the home. One can use the amortization formula, $A = d\left[\dfrac{1-(1+i)^{-n}}{i}\right]$, to determine the amount of equity given the number of payments originally required, the number of payments made, the amount of each payment, the original loan amount, and the interest per payment period. Here, n is the number of payments left to be made on the life of the loan (number of payments originally required minus the number of payments made).

✎ Question 4

A couple buys a house for $250,000 at 4.5% interest. After making payments for a year and a half, they decide to sell their home. If the original mortgage was for 30 years, how much equity do they have? Round to the nearest $100.

Answer

$6100

Section 22.4   Annuities

ᒉᔆ Example I

An annuity has a value of $50,000 and will be paid in equal monthly payments over 30 years at 6% annual interest. How much would each monthly payment be?

Solution

Because the annual rate is 6%, $i = \frac{0.06}{12} = 0.005$. Set up the annuity formula with $n = 12 \times 30 = 360$.

$$d = \frac{Ai}{1-(1+i)^{-n}} = \frac{\$50,000 \times 0.005}{1-(1+0.005)^{-360}} = \$299.78$$

Homework Help

Exercises 1 – 8
Carefully read Section 22.1 before responding to these exercises.

Exercises 9 – 14
Carefully read Section 22.2 before responding to these exercises.

Exercises 15 – 42
Carefully read Section 22.3 before responding to these exercises.

Exercises 43 – 47
Carefully read Section 22.4 before responding to these exercises.

Do You Know the Terms?

Cut out the following 19 flashcards to test yourself on Review Vocabulary. You can also find these flashcards at http://www.whfreeman.com/fapp7e.

Chapter 22 Borrowing Models **Add-on loan**	Chapter 22 Borrowing Models **Adjustable rate mortgage (ARM)**
Chapter 22 Borrowing Models **Amortization formula**	Chapter 22 Borrowing Models **Amortize**
Chapter 22 Borrowing Models **Annual percentage rate (APR)**	Chapter 22 Borrowing Models **Annuity**
Chapter 22 Borrowing Models **Compound interest formula**	Chapter 22 Borrowing Models **Compounding period**

One whose interest rate can vary during the course of a loan.	A loan in which you borrow the principal and pay back principal plus total interest with equal payments.
To repay in regular installments.	Formula for installment loans that relates the principal A, the interest rate i per compounding period, the payment d at the end of each period, and the number of compounding periods n needed to pay off the loan: $A = d\left[\dfrac{1-(1+i)^{-n}}{i}\right]$.
A specified number of (usually equal) payments at equal intervals of time.	The rate of interest per compounding period times the number of compounding periods per year.
The fundamental interval for compounding, within which no compounding is done.	Formula for the amount in an account that pays compound interest periodically. For an initial principal P and effective rate i per compounding period, the amount after n compounding periods is $A = P(1+i)^n$.

Chapter 22 Borrowing Models **Conventional loan**	Chapter 22 Borrowing Models **Discounted loan**
Chapter 22 Borrowing Models **Effective annual rate (EAR)**	Chapter 22 Borrowing Models **Effective rate**
Chapter 22 Borrowing Models **Equity**	Chapter 22 Borrowing Models **Interest**
Chapter 22 Borrowing Models **Nominal rate**	Chapter 22 Borrowing Models **Principal**

A loan in which you borrow the principal minus the interest but pay back the entire principal with equal amounts.	A loan in which each payment pays all the current interest and also repays part of the principal.
The actual percentage rate, taking into account compounding.	The effective rate per year.
Money earned on a loan.	The amount of principal of a loan that has been repaid.
Initial balance.	A stated rate of interest for a specified length of time; a nominal rate does not take into account any compounding.

**Chapter 22
Borrowing Models**

Savings formula

**Chapter 22
Borrowing Models**

Simple interest

The method of paying interest on only the initial balance in an account and not on any accrued interest.   For a principal P, an interest rate r per year, and t years, the interest I is $I = Prt$.

Formula for the amount in an account to which a regular deposit is made (equal for each period) and interest is credited, both at the end of each period. For a regular deposit of d and an interest rate i per compounding period, the amount A accumulated is

$$A = d\left[\frac{(1+i)^n - 1}{i}\right].$$

Practice Quiz

1. A 30-year U.S. Treasury bond with a yield of 4.91% was issued on February 15, 2001, for $10,000. How much interest would the Treasury pay on it through February 15, 2031?

    a. $1,473

    b. $24,730

    c. $14,730

2. Suppose you need to borrow $2,752 to pay for your Fall 2006 tuition. The credit union offers you a 6% add-on loan to be repaid in quarterly installments over two years. What is your quarterly payment?

    a. $385.28

    b. $770.56

    c. $128.43

3. Suppose that you owe $1000 on your credit card, which charges 0.8325% interest per month, and you just let the balance ride. After 12 months of letting your balance ride (neglecting any finance charges), your new balance is

    a. $1008.36

    b. $2610.62

    c. $1104.60

4. You are required to pay back a loan of $2275. You work three months in the summer and each month deposit an amount, d, into a savings account that pays 4.5% per year. How much money do you need to deposit each month to have enough to pay off the loan at the end of the summer?

    a. $725.21

    b. $755.50

    c. $565.56

5. A platinum credit card is currently offering a fixed introductory annual percentage rate (APR) of 2.99% on purchases and balance transfers. Find the effective annual rate (EAR).

    a. 3.03%

    b. 1.002%

    c. 4.24%

6. If you buy a home by taking a 30-year mortgage for $80,000, and an interest rate of 8% compounded monthly, how much will the monthly payments be?

    a. about $805

    b. about $587

    c. about $640

7. The Hochwalds purchased a home in 1984 for $110,000. They made a down payment of $50,000 and financed the balance with a 30-year mortgage at an $8\frac{3}{4}$% interest rate. How much were their monthly payments?

    a. $393.35

    b. $674.40

    c. $472.02

8. Using the information from Problem 7, how much equity did the Hochwalds have in their home after exactly 18 years? (Assume the house is still worth $110,000.)

    a. $91,994.78

    b. $68,005.22

    c. $41,994.78

9. Which type of annuity best describes the following arrangement: equal payments over the life of the annuity?

    I. Ordinary annuity

    II. Life Income annuity

    a. I only

    b. II only

    c. both I and II

10. Which statement is true?

    I. Monthly annuity payments to men and women are approximately equal.

    II. Monthly annuity payments to men are generally higher than monthly annuity payments to women.

    a. I only

    b. II only

    c. neither I nor II

Word Search

Refer to page 850 of your text to obtain the Review Vocabulary. There are 18 hidden vocabulary words/expressions in the word search below. All vocabulary words/expressions are represented in the word search. It should be noted that spaces are removed as well as hyphens. Also, abbreviations do not appear in the word search.

```
Y I R F S Z I X S V Y E Y H I T R E T Q P C A I M
I C G A R A E F F E C T I V E R A T E Q S G N Z N
E K O E R O T H N S E I R F F Y T K O A M E Q M L
T N M I R S A V I N G S F O R M U L A E D E X Y E
E Z I I B R R L N H A T O C M S G I N E A K W O M
T A R D Q S E W T T G H D L E N A O L N O D D A T
A M B O F F G J E P T L T L Z R E P R E F U T I A
R O T I S Q A C R O R P N E A E V D L V E O S Q A
L R S R A R T I E F O T H I R P D I I R M H E O N
A T E E R G N V S C M S S S B I I S E I L E N D N
U I R P B I E E T O E V P A S K E C N X G E H P U
N Z E G G E C L S N T T N D U E J O N G C R E U I
N A T N H N R P H V A V O E W N B U X I S P E G T
A T N I K C E Z R E R C M X S H R N S T R O E I Y
E I I D Q T P D L N E N I E E E O T X O M P G U H
V O E N E T L E N T L E N O E K F E S T G Q O A E
I N L U M E A N S I B T A M I X E D R M R E T K Q
T F P O U H U E S O A R L A F G N L Z W R G E R U
C O M P O U N D I N T E R E S T F O R M U L A N I
E R I M D R N F M A S H A N Y S I A D O I R E P T
F M S O F A A T D L U C T S E M G N Y M F E O I Y
F U N C W V S D V L J J E N F R Q E D N A E O J D
E L T E N R S N S O D V A A I T Y E U Z D C E G M
X A R D S O U N S A A R S E P D E E Z I T R O M A
P S G K P O G L L N F W I D E N X C L N S E J S G
```

1. _____   10. _____

2. _____   11. _____

3. _____   12. _____

4. _____   13. _____

5. _____   14. _____

6. _____   15. _____

7. _____   16. _____

8. _____   17. _____

9. _____   18. _____

Chapter 23
The Economics of Resources

Chapter Objectives

Check off these skills when you feel that you have mastered them.

☐ Track the changing rate of a population growing geometrically at a fixed rate.

☐ Understand the meaning of a population's carrying capacity.

☐ Understand the difference between the static reserve of a nonrenewable resource and its exponential reserve, and calculate the exponential reserve from the formula.

☐ Understand the meaning of a reproduction curve, and explain why a sustainable yield policy is needed for a harvestable resource.

☐ Estimate the maximum sustainable yield for a harvestable resource from its reproduction curve.

☐ Describe chaos and the butterfly effect.

☐ Use the logistic population model to calculate the fraction of a population after a period of time.

☐ Explain the relationship of chaos to the logistic population model.

Guided Reading

Introduction

As individuals and as a society, we are faced with the need to find a balance between using resources and conserving them. Some of these resources replenish themselves but some are nonrenewable. This chapter explores models for the decay or depletion of resources.

Section 23.1 Growth Models for Biological Populations

 **Key idea**

If we use a geometric growth model to describe and predict human (or other species) populations, the effective rate of growth is the difference between population increase caused by births and decrease caused by deaths. This difference is called the **rate of natural increase**.

ᴳᵥ Example A

To predict the population of the US in the year 2010, we can use the following census information from 1990.

$$P = 255 \text{ million}, r = 0.7\% = 0.007.$$

With annual compounding, we can predict that the population 20 years later, in 2010 would be the following.

$$\text{Population in } 2010 = 255 \text{ million} \times (1 + 0.007)^{20} = 293 \text{ million}$$

At this rate, the population would double in about 100 years.

☞ Key idea

Small changes in the birth or death rate will affect the rate of natural increase, and this changes our prediction significantly.

ᴳᵥ Example B

Consider three countries A, B, and C, each of whose population in 1995 was 120 million. Country A is growing at a rate of 0.5%, country B at 1%, and country C at 2%.

a)   What would you predict the population of each country to be in 2010?

b)   What would you predict for 2025?

Solution

a)   The year 2010 is 15 years later than 1995, so the following formula for $r = 0.005$, 0.01, and 0.02 for countries A, B, and C, respectively, as follows.

$$\text{Population in } 2010 = 120 \text{ million} \times (1 + r)^{15}$$

b)   The year 2025 is 30 years later than 1995 so the following formula for $r = 0.005$, 0.01, and 0.02 for countries A, B, and C, respectively, as follows.

$$\text{Population in } 2010 = 120 \text{ million} \times (1 + r)^{30}$$

The table of population values for the three countries looks like the following.

Country	1995	2010	2025
A	120 million	129 million	139 million
B	120 million	139 million	162 million
C	120 million	162 million	217 million

☞ Key idea

A population cannot keep growing without limit. The resources available to the population limit the size of that population. A population limit in a particular environment is called the **carrying capacity**.

☞ Key idea

The closer a population gets to its carrying capacity, the more slowly the population will grow. The logistics model for population growth takes carrying capacity into account by reducing the annual increase of rP by a factor of how close the population size P is to the carrying capacity M. It is given by the following.

$$\text{growth rate } P' = rP\left(1 - \frac{\text{population size}}{\text{carrying capacity}}\right) = rP\left(1 - \frac{P}{M}\right)$$

⌒ Example C

A salmon fishery has a carrying capacity of 100,000 fish. The natural rate of increase for the population is 3% per year. What is the growth rate of the population if the population is at 52,000 fish?

Solution

growth rate $= 0.03\left(1 - \frac{52,000}{100,000}\right) = 0.0144$, 1.44% per year.

Section 23.2 How Long Can a Nonrenewable Resource Last?

⊶ Key idea

A **nonrenewable resource**, such as a fossil fuel or a mineral ore deposit, is a natural resource that does not replenish itself.

⊶ Key idea

A growing population is likely to use a nonrenewable resource at an increasing rate. The regular and increasing withdrawals from the resource pool are analogous to regular deposits in a sinking fund with interest, and the same formula applies to calculate the accumulated amount of the resource that has been used, and is thus gone forever. The **static reserve** is the time the resource will last with constant use; the **exponential reserve** is the time it will last with use increasing geometrically with the population.

⊶ Key idea

The formula for the exponential reserve of a resource with supply S, initial annual use U, and usage growth rate r is as follows.

$$n = \frac{\ln\left[1 + \left(\frac{S}{U}\right)r\right]}{\ln\left[1 + r\right]}$$

Here, ln is the natural logarithm function, available on your calculator.

⌒ Example D

Imagine that a certain iron ore deposit will last for 200 years at the current usage rate. How long would that same deposit last if usage increases at the rate of 4% each year?

Solution

The static reserve, $\frac{S}{U}$, is 200 years, so we can plug that value into the formula to solve for n, using

the assumed 4% = .04 for r. We obtain $n = \dfrac{\ln\left[1 + 200(0.04)\right]}{\ln\left[1 + 0.04\right]} = \dfrac{\ln 9}{\ln 1.04} \approx 56$ years.

✎ Question 1

In 2005, the Acme corporation had non-renewable resources of 3479 million pounds of materials. The annual consumption was 98 million pounds. The projected company consumption will increase 2.3% per year, through 2020.

a) What is the static reserve in 2005?

b) What is the exponential reserve in 2005?

Answer

a) 35.5 years

b) 26.25 years

Section 23.3   Sustaining Renewable Resources

☞ Key idea

A **renewable natural resource** replenishes itself at a natural rate and can often be harvested at moderate levels for economic or social purposes without damaging its regrowth.   Since heavy harvesting may overwhelm and destroy the population, economics and conservation are crucial ingredients in formulating proper harvesting policies.

☞ Key idea

We keep track of the population (measured in biomass) from one year to the next using a **reproduction curve**. Under normal conditions, natural reproduction will produce a geometrically growing population, but too high a population level is likely to lead to overcrowding and to strain the available resources, thus resulting in a population decrease. This model leads to a reproduction curve looking something like this:

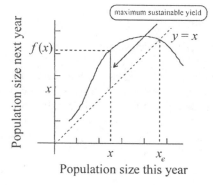

☞ Key idea

The dotted 45° diagonal line is the set of points where the population would be unchanged from year to year, and any point where it intersects the reproduction curve is an **equilibrium population size**.

☞ Key idea

The marked population value x is the level which produces maximum natural increase or yield in a year, and the difference between x and $f(x)$ (the population level a year later) is the maximum sustainable yield (or harvest) $f(x) - x$. This amount is the maximum that may be harvested each year without damaging the population, and represents a good choice for a sustained-yield harvesting policy.

Section 23.4   The Economics of Harvesting Resources

☞ Key idea

If our main concern is profit, we must take into account the economic value of our harvest and the cost of harvesting. If we also include in our model **economy of scale** (denser populations are easier to harvest), then the sustainable harvest which yields a maximum profit may be smaller than the maximum sustainable yield.

☞ Key idea

Finally, if we also take into account the economic value of capital and consider profit as our only motivation, it may be most profitable to harvest the entire population, effectively killing it, and invest the profits elsewhere. The history of the lumbering and fishing industries demonstrates this unfortunate fact.

Section 23.5   Dynamical Systems and Chaos

☞ Key idea
In some populations, the state of the population depends only on its state at previous times. This kind of system is called a **dynamical system**. For example, a population's size in a given year may depend entirely upon its size in the previous year.

☞ Key idea
Behavior that is determined by preceding events but is unpredictable in the long run is called **chaos**.

☞ Key idea
In some systems a small change in the initial conditions can make a huge difference later on. This is the **butterfly effect.**

☞ Key idea
The logistic population model can illustrate chaos in biological population. Consider the current year's population as a fraction x of the carrying capacity, and next year's population as a fraction $f(x)$. The amount by which the population is multiplied each year is $\lambda = 1 + r$, where r is the population's annual growth rate. Then the logistic model can be written as $f(x) = \lambda x(1-x)$.

↶ Example F
A population grows according to a logistic growth model, with population parameter $\lambda = 1.4$ and $x = 0.75$ for the first year. What is the next population fraction?

Solution
$1.4(0.75)(1-0.75) = 0.2625$

☞ Key idea
The logistic model illustrates chaotic behavior when the population parameter λ is equal to 4. In this case, for any starting population fraction, the population fraction changes year after year in no predictable pattern.

✐ Question 2
Let $f(n) = $ the sum of the cubes of the digits of n. Start with 371 and apply f repeatedly. Start with 234 and apply f repeatedly. Start with 313 and apply f repeatedly. Are the three behaviors the same?

Answer
no

Homework Help

Exercises 1 – 10
Carefully read Section 23.1 before responding to these exercises.

Exercises 11 – 20
Carefully read Section 23.2 before responding to these exercises.

Exercises 21 – 38
Carefully read Sections 23.3 and 23.4 before responding to these exercises.

Exercises 39 – 45
Carefully read Section 23.5 before responding to these exercises.

Do You Know the Terms?

Cut out the following 25 flashcards to test yourself on Review Vocabulary. You can also find these flashcards at http://www.whfreeman.com/fapp7e.

Chapter 23 The Economics of Resources **Biomass**	Chapter 23 The Economics of Resources **Butterfly effect**
Chapter 23 The Economics of Resources **Carrying capacity**	Chapter 23 The Economics of Resources **Chaos**
Chapter 23 The Economics of Resources **Cobweb design**	Chapter 23 The Economics of Resources **Compound interest formula**
Chapter 23 The Economics of Resources **Deterministic**	Chapter 23 The Economics of Resources **Dynamical system**

A small change in initial conditions of a system can make an enormous difference later on.

A measure of a population in common units of equal value.

Complex but deterministic behavior that is unpredictable in the long run.

The maximum population size that can be supported by the available resources.

Formula for the amount in an account that pays compound interest periodically. For an initial principal P and effective rate r per year, the amount after n years is $A = P(1+r)^n$.

A kind of graphical portrayal of the evolution of a dynamical system, such as a population.

A system whose state depends only on its states at previous times.

A system is this if its future behavior is completely determined by its present state, past history, and known laws.

Chapter 23 The Economics of Resources **Economy of scale**	Chapter 23 The Economics of Resources **Equilibrium population size**
Chapter 23 The Economics of Resources **Exponential reserve**	Chapter 23 The Economics of Resources **Iterated function system (IFS)**
Chapter 23 The Economics of Resources **Logistic model**	Chapter 23 The Economics of Resources **Maximum sustainable yield**
Chapter 23 The Economics of Resources **Natural increase**	Chapter 23 The Economics of Resources **Nonrenewable resource**

A population size that does not change from year to year.	Costs per unit decrease with increasing volume.
A sequence of elements (number or geometric shapes) in which the next element is produced from the previous one according to a function (rule).	How long a fixed amount of a resource will last at a constantly increasing rate of use. A supply S, as an initial rate of use U that is increasing by a proportion r each year, will last $\dfrac{\ln\left(1+\dfrac{S}{U}r\right)}{\ln(1+r)}$ years.
The largest harvest that can be repeated indefinitely.	A particular population model that begins with near-geometric growth but then tapers off toward a limiting population (the carrying capacity).
A resource that does not tend to replenish itself.	The growth of a population that is not harvested.

Chapter 23 The Economics of Resources **Population structure**	Chapter 23 The Economics of Resources **Rate of natural increase**
Chapter 23 The Economics of Resources **Renewable natural resource**	Chapter 23 The Economics of Resources **Reproduction curve**
Chapter 23 The Economics of Resources **Savings formula**	Chapter 23 The Economics of Resources **Static reserve**
Chapter 23 The Economics of Resources **Sustainable yield**	Chapter 23 The Economics of Resources **Sustained-yield harvesting policy**

Birth rate minus death rate; the annual rate of population growth without taking into account net migration.	The division of a population into subgroups.
A curve that shows population size in the next year plotted against population size in the current year.	A resource that tends to replenish itself; examples are fish, forests, wildlife.
How long a fixed amount of a resource will last at a constant rate of use; a supply S used at an annual rate U will last S/U years.	Formula for the amount in an account to which a regular deposit is made (equal for each period) and interest is credited, both at the end of each period. For a regular deposit of d and an interest rate i per compounding period, the amount A accumulated is $A = d\left[\dfrac{(1+i)^n - 1}{i}\right]$.
A harvesting policy that can be continued indefinitely while maintaining the same yield.	A harvest that can be continued at the same level indefinitely.

Chapter 23
The Economics of Resources

Yield

The amount harvested at each harvest.

Practice Quiz

1. The Italian population was 47,105 in 1950. This population was said to be increasing at an average growth rate of 0.6% per year. In 1960, the Italian census reported a population of 50,198. This number represents _____.

   a. substantially more than the expected population growth for the 10-year period.

   b. substantially less than the expected population growth for the 10-year period.

   c. approximately the expected population growth for the 10-year period.

2. Which of the following is true?

   a. An equilibrium population size changes from year to year.

   b. In the logistic model, if the population exceeds the carrying capacity, the population decreases.

   c. The static reserve is how long a supply will last at an initial rate of use that is increasing by a proportion each year.

3. Production of oil exploration peaked in 1962. The remaining reserves are expected to last another 44 years at the current rate of use. Approximately how long would the supply last if the rate of use increased 3% per year?

   a. 44 years

   b. 9 years

   c. 28 years

4. Imagine that a certain iron ore deposit will last for 400 years at the current usage rate. How long would that same deposit last if usage increases at the rate of 5% each year?

   a. 53 years

   b. 62 years

   c. It never will be depleted.

5. For the following model, $f(x) = 4x(2-x)$, if the starting population fraction is 0.1, what is the next population fraction?

   a. 0.76

   b. 0.01

   c. 0.41

6. Given the reproduction curve below

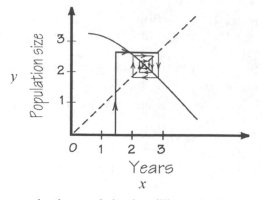

where x represents years and y the population in millions, the dynamics, over time, for an initial starting population of approximately 1.3 million would indicate _____.

a. after an initial adjustment, the population goes to the equilibrium population and stays there.

b. after an initial adjustment, the population cycles between values over and under the equilibrium population.

c. the population spirals in toward the equilibrium population.

7. An optimal harvesting policy for renewable resources depends on _____.

a. price

b. cost

c. both a and b

8. You notice that the coworkers in your office complex have not been looking too happy of late. You decide to go into work the next day and smile at everyone you see. By the end of the work day you notice that this has created a rippling effect and most of your coworkers are smiling. This phenomenon is an example of _____.

a. chaos

b. the butterfly effect

c. a dynamical system

9. Maximum Sustainable Yield (MSY) has become a common practice in fisheries management. MSY is dependent on _____.

I. growth rates and over-fishing.

II. birth rates and mortality rates.

a. I only

b. II only

c. both I and II

10. Which of the following statements is true?

I. Chaotic behavior is not predictable in the short term.

II. Chaotic behavior is predictable in the long term.

a. I only

b. II only

c. neither I nor II

Word Search

Refer to pages 885 – 886 of your text to obtain the Review Vocabulary. There are 25 hidden vocabulary words/expressions in the word search below. This represents all 25 words/expressions in the Review Vocabulary. It should be noted that spaces and hyphens, are removed as well as apostrophes. Also, the abbreviations do not appear in the word search. The backside of this page has additional space for the words/expressions that you find.

```
F T F E B F N S L S M V T D D E E T A R L A N I M O N K S E
O I A Z M T S S E F A Q C O B W E B D E S I G N Y W R T Z S
E P S I C O D P S F X A N H E E F W N P E J D I L D A N W T
N P O S C E X D M P I I B A J N O R F R G U E E A S T S E A
E E R N N T I T R O M S V V R E O P A O M L C X R G E E E T
M H N O E A M E C P U D E O E M G G B D D E R P K W O L A I
I S H I P C E E M U M X C I N Y V N U U D O U O T T F O U C
Q C A T V O S C A L S H D T E P N M T C F N O N V S N G J R
L C E A P M O O T A U M S E W D B D T T D E S E E O A I E E
Y C I L O P G N I T S E V R A H D L E I Y D E N I A T S U S
T E L U P O F O E I T E I A B T E A R O N S R T I T U T L E
I S C P U U T M E O A G H T L T T H F N A A E I O G R I P R
C S M O L N E Y E N I N P E E T E J L C M D L A P Q A C M V
A Q Q P A D P O E G N O G D N T R S Y U I A B L C M L M B E
P N X M T I P F B R A P C F A N M J E R C A A R X E I O I B
A R A U I N I S T O B S E U T E I A F V A C W E K S N D E D
C F C I O T T C C W L A C N U L N L F E L W E S T H C E H E
G S G R N E D A P T E T I C R E I U E K S B N E E Y R L R E
N G S B S R I L R H Y F C T A V S M C N Y G E R H L E Q J M
I S A I T E M E L H I C N I L T T R T V S C R V O D A D R C
Y D R L R S S W R R E I E O R O I O R L T D N E P N S R H H
R A Q I U T E E A E L P N N E K C F A I E A O J R Q E C S I
R G P U C F O V S O D G T S S S X S S G M U N E I E N C O P
A H N Q T O M A T D Q E E Y O S M G S S A M O I B E S Y R E
C T M E U R F A T D S H E S U E I N A L O S E F Y O O I S O
B I S R R M G D M O D A E T R H K I U M D R F F I B A S O S
W P A H E U H R A N I D U E C E L V E I Q R V P E E H O R E
A S V V S L I K U T O T A M E Y B A L A S I B M G V C S T E
D H O K N A T U R A L I N C R E A S E M N N R I P R L F E N
N P A E I S M D L E I Y E L B A N I A T S U S P I I G O L C
```

1. _____   7. _____

2. _____   8. _____

3. _____   9. _____

4. _____   10. _____

5. _____   11. _____

6. _____   12. _____

13. _____

14. _____

15. _____

16. _____

17. _____

18. _____

19. _____

20. _____

21. _____

22. _____

23. _____

24. _____

25. _____

Practice Quiz Answers

Chapter 1	1. b	2. b	3. b	4. c	5. c	6. c	7. a	8. a	9. b	10. c
Chapter 2	1. c	2. b	3. b	4. b	5. c	6. a	7. c	8. a	9. c	10. b
Chapter 3	1. c	2. c	3. a	4. b	5. b	6. c	7. c	8. a	9. c	10. c
Chapter 4	1. a	2. b	3. c	4. a	5. b	6. b	7. c	8. c	9. c	10. a
Chapter 5	1. a	2. b	3. b	4. c	5. c	6. b	7. c	8. b	9. a	10. c
Chapter 6	1. c	2. b	3. b	4. b	5. a	6. b	7. b	8. a	9. c	10. c
Chapter 7	1. b	2. c	3. a	4. b	5. b	6. b	7. a	8. a	9. a	10. a
Chapter 8	1. c	2. a	3. a	4. b	5. c	6. b	7. a	8. c	9. a	10. b
Chapter 9	1. a	2. c	3. c	4. a	5. b	6. a	7. c	8. b	9. c	10. c
Chapter 10	1. a	2. b	3. a	4. b	5. c	6. a	7. c	8. b	9. c	10. b
Chapter 11	1. b	2. c	3. b	4. b	5. b	6. c	7. a	8. b	9. a	10. a
Chapter 12	1. b	2. c	3. a	4. c	5. a	6. b	7. c	8. c	9. b	10. c
Chapter 13	1. c	2. b	3. b	4. c	5. b	6. a	7. a	8. b	9. b	10. b
Chapter 14	1. c	2. c	3. a	4. c	5. b	6. c	7. b	8. b	9. c	10. c
Chapter 15	1. b	2. b	3. c	4. b	5. b	6. c	7. a	8. a	9. a	10. a
Chapter 16	1. a	2. a	3. a	4. c	5. c	6. c	7. a	8. a	9. a	10. a
Chapter 17	1. a	2. b	3. c	4. b	5. c	6. b	7. b	8. c	9. a	10. a
Chapter 18	1. a	2. c	3. b	4. b	5. a	6. b	7. c	8. b	9. c	10. a
Chapter 19	1. b	2. b	3. c	4. a	5. c	6. a	7. a	8. a	9. b	10. c
Chapter 20	1. a	2. b	3. a	4. c	5. c	6. b	7. c	8. b	9. b	10. c
Chapter 21	1. c	2. a	3. b	4. b	5. a	6. c	7. a	8. c	9. c	10. b
Chapter 22	1. c	2. a	3. c	4. b	5. a	6. b	7. c	8. b	9. a	10. b
Chapter 23	1. c	2. b	3. c	4. b	5. a	6. c	7. c	8. b	9. c	10. c